A Guide to...
Cockatiels
and their Mutations
as Pet and Aviary Birds

By Dr Terry Martin BVSc and Diana Andersen

Edited and Published by ABK Publications © 2007

© **ABK Publications 2007**

**ABK Publications
PO Box 6288,
South Tweed Heads,
NSW 2486 Australia.**

ISBN 978 0 9750817 8 5 (hard cover)

ISBN 978 0 9750817 7 8 (soft cover)

All rights reserved. No part of this publication may be reproduced, stored in any retrieval system, or transmitted in any form or by any means without the prior permission in writing of the publisher.

Disclaimer: Very few drugs are registered for use in birds, and most usages and dose rates have been extrapolated from mammalian therapeutics. Everyone using medications should be aware that manufacturers of these drugs will not accept any responsibility for the 'off-label' use of their drugs. The dose rates and information are based on clinical trials and practical experience, but unrecorded adverse side effects may occur. Where possible, the author has provided brand names for the drugs mentioned. These should not be taken as a recommendation for one particular brand over another, but rather as a starting point for you to find the drug of your choice. In most instances, contraindications and side effects are not listed. This should not be taken to mean that there are none—many of these drugs have not been used extensively, and reports on contraindications and side effects are not recorded at date of publication.

Front Cover:
Top left: Normal Grey Cockatiel cock — Peter Odekerken
Top right: Yellowface Cinnamon Pearl Cockatiel hen — Nancy Johnson-Mello
Bottom left: Whiteface Dilute Grey Cockatiel hen — Peter Odekerken
Bottom right: Paleface Cinnamon Pied Cockatiel cock — Peter Odekerken

Back Cover:
Paleface Grey Cockatiel cock — Nancy Johnson-Mello

Design, Type and Art: PrintHouse Multimedia Graphics

CONTENTS

ABOUT THE AUTHORS ... 9
Contributing Authors ... 10
ACKNOWLEDGEMENTS ... 11
INTRODUCTION ... 12
Other Names ... 13
Description ... 14
Cock, Hen, Immature
Voice ... 15
COCKATIELS IN THE WILD ... 16
Origins ... 16
Present Distribution ... 17
Food and Feeding Habits ... 18
Breeding ... 20
Status ... 20
COCKATIELS IN CAPTIVITY ... 21
AVIARY BIRDS ... 22
History ... 22
Show Standard ... 22
Behaviour ... 23
Acquiring Stock ... 23
Quarantine ... 26
HOUSING ... 27
Conventional Aviaries ... 27
Suspended Aviaries ... 28
Aviary Materials ... 30
Wire
Shelter and Insulation ... 30
Walkways ... 31
Perches and Other Aviary Furnishings ... 32
Nestboxes ... 32
Substrate ... 34
Rodent Control ... 34
FEEDING ... 35
Nutrition ... 35
Pellets ... 37
Converting Your Birds to a Pelleted Diet, Case Study
Seeds ... 38
Sprouted Seed ... 39
Sprouting Method
Fruits and Vegetables ... 40
Native Foods ... 40
Seeding Grasses ... 41
Persuading Your Birds to Eat New Foods ... 41
Water ... 41

Page 3

BREEDING ... 42
Selection of Breeding Stock ... 42
Gender Identification ... 42
   Colour Mutations
Breeding Season ... 45
Colony Breeding ... 46
Nestbox Preparation ... 46
Courtship ... 47
Egg Laying and Nest Inspection ... 47
Incubation ... 48
Incubation Problems ... 48
   Infertile or Clear Eggs, Hens That Fail to Incubate,
   Cocks That Fail to Incubate
PARENT REARING ... 49
Chick Development ... 50
   Monitoring Growth, Physical Check,
   Feather Development, Fledging
Feather Plucking ... 53
ARTIFICIAL INCUBATION ... 54
Hatching ... 55
Brooding Conditions ... 56
   Homemade Brooders, Humidity Levels
Maintaining Hygiene ... 58
HANDREARING ... 58
Assessing the Condition of Chicks ... 58
Handrearing Formulas ... 59
Feeding Instruments ... 61
Feeding Methods ... 61
Feeding Regime ... 63
Monitoring Chick Development ... 63
   Weight Charts
Body Condition ... 65
Weaning ... 65
   Weaning Cage
BASIC FIRST AID FOR BIRDS ... 67
Hospital Cage Accessories ... 68
Hospital Cage Set-up ... 68
Common Injuries ... 68
COMPANION BIRDS ... 69
Choosing a Pet Bird ... 70
One Pet Cockatiel or Two? ... 72
Veterinary Check ... 73
Preparing a Home for Your Pet Cockatiel ... 73
   Perches, Food and Water Dishes, The Cage Floor,
   Cage Location
The First Days at Home ... 76

Diet	77
Enrichment	77
Toys, Natural Branches, Flight, Basic Behaviour Training, Bathing and Spraying	
Talking	80
Household Hazards	80
Wing Clipping	80
Should Wing Clipping be Done?, How Should a Wing Clip be Done?, Blood Feathers, Wing Clip Patterns, Who Should Perform This Procedure?, How Long Does a Clip Last?	
Restraint Alternatives	83
Avoiding Behavioural Problems	83
Selecting the Right Cockatiel, Food for Thought, The Environment, Lifelong Education, Ten Ways to Success	

COLOUR MUTATIONS AND GENETICS — 87

COLOUR BREEDING — 88

Identification of Breeding Stock and Offspring	88
Banding	
Accurate Record Keeping	89
Understanding Genetics	89

BASIC GENETICS — 91

Autosomal Recessive	91
Sample Pairings	
Calculating Multi-Autosomal Mutations	94
Autosomal Co-dominant	95
Sex Chromosomes in Birds	96
Sex-Linked Recessive	96
Sample Pairings	
Calculating Sex-linked Mutations	99
Calculating Combinations between Sex-linked and Autosomal Mutations	101

MUTATIONS — 103

| Wildtype Pigmentation | 103 |
| Mutation versus Colour | 104 |

ESTABLISHED MUTATIONS — 105

Melanin-altering Mutations	105
Albinistic Genes, Dilution Genes, Leucistic Genes	
Psittacin-altering Mutations	105
Pattern Mutations	105

UNIVERSAL (WORLDWIDE) MUTATIONS — 106

Sex-linked Lutino (Z^{ino})	106
Desirable Matings	
Cinnamon (Z^{cin})	108
Desirable Matings	
Recessive Pied (r)	110
Desirable Matings	

Whiteface (b)	115
Desirable Matings	
Paleface ('Pastelface') (b^{aq})	117
Desirable Matings	
Opaline (Pearl) (Z^{op})	119
Desirable Matings	
Yellow Suffusion	122
Desirable Matings	
Silver	123
REGIONAL MUTATIONS	126
European Mutations	126
Dominant Edged (Dominant Silver) (E)	126
Desirable Matings	
Ashen Fallow ('Recessive Silver') (ash)	128
Desirable Matings	
Sex-linked Yellowcheek (Z^{yc})	130
Desirable Matings	
Non Sex-linked Lutino (a)	131
North American Mutations	132
Bronze Fallow (a^{bz})	132
Desirable Matings	
Dominant Yellowface (T)	133
Desirable Matings	
'Emerald'	134
Desirable Matings	
'Goldcheek'	136
Australian Mutations	137
Faded ('West Coast Silver') (fd)	137
Desirable Matings	
Platinum (Z ino^{pl})	139
Desirable Matings	
Dilute ('Pastel Silver') (dil^{gw})	143
Desirable Matings	
Australian Fallow (f)	145
Desirable Matings	
Edged Dilute ('Silver Spangle') (ed)	147
Desirable Matings	
Suffused ('Olive') (dil)	149
Desirable Matings	
Pewter (Z^{pw})	151
Desirable Matings	
Australian 'Yellowface'	153
A New Australian Mutation?	
COMBINATION COLOURS	154
Melanin Combinations	154
Pearl Combinations	156

Pied Combinations	158
Pearl Pied Combinations	159
Whiteface Combinations	160
Whiteface Pearl Combinations	164
Whiteface Pied Combinations	165
Whiteface Pearl Pied Combinations	166
Paleface Combinations	167
Yellowface Combinations	171
Yellowcheek Combinations	172
COLOUR ODDITIES	174
'Halfsiders'	174
Schimmel	175
Acquired Colours	175
Cockatiel Mutation Technical and Common Names	177
HEALTH AND DISEASE	178
Disclaimer	179
General Information	179
How to Use This Chapter	179
Supply of Prescription Medications	180
PART ONE	180
Change in Droppings	180
Appetite and Thirst	181
Posture	181
Feathers and Skin	183
Wings	184
Feet and Legs	184
Head	185
Physical Examination	187
Paediatrics	188
PART TWO	189
Infectious Diseases	189
Psittacosis, Gastrointestinal Infections, Intestinal Parasites, Megabacteria, Respiratory Infections, Psittacine Beak and Feather Disease (PBFD) and Avian Polyomavirus (APV)	
Non-infectious Diseases	191
Malnutrition, Kidney Disease, Liver Disease, Zinc and Lead Poisoning, Reproductive Disease, Diabetes, Pancreatic Disease, Paediatric Problems	
BIBLIOGRAPHY	194

Normal Grey Cockatiel pair (cock on right).

ABOUT THE AUTHORS

Dr Terry Martin BVSc

Avicultural geneticist, Dr Terry Martin prepared the Mutations chapters. Terry began keeping birds at 10 years of age when he was given a pair of Zebra Finches. Colour mutations quickly became his primary interest in this and other species, which led to a solid knowledge of basic genetics at an early age. In 1988 Terry graduated with a degree in Veterinary Science from the University of Queensland.

Terry has dedicated his ongoing studies of birds to the area of genetics, pigmentation and colour mutations and is highly respected in his research field internationally. He has presented at avian and veterinary meetings and conferences, contributed articles to various publications including **Australian BirdKeeper Magazine** and various **ABK Publications** titles and authored the book **A Guide to Colour Mutations and Genetics in Parrots** published by **ABK Publications**.

In 1999 Terry initiated the formation of an international genetics discussion group. Since that time he has continuously discussed and studied mutations and genetics of all avicultural species with authorities from around the world.

He has been a member of the Australian National Cockatiel Society since 1995 and, although time only allows his occasional attendance at meetings, he has given presentations on genetics on a number of occasions.

Diana Andersen

Diana Andersen, together with Peggy Cross, co-authored the first edition of **A Guide to Cockatiels and their Mutations—their Management, Care and Breeding**. Diana is renowned for her thorough knowledge of many species of animals—both bird and mammal. After attaining a Bachelor of Arts degree, majoring in Design in 1981, she became a jewellery maker. Following this period Diana spent three years at the Cohuna Koala Park where she was responsible for the feeding and care of the park's bird and mammal collection.

In 1999, Diana was appointed bird keeper within the Australian Fauna Department of the Perth Zoo—responsible for all aspects of management and care of all avian species, as well as some of the Australian mammals at the zoo. Her responsibilities and experience over the next seven years at the zoo included working on the Native Species Breeding Programs for critically endangered Australian species.

Alongside her busy work schedule Diana has maintained an active participation in all aspects of aviculture. She has presented her knowledge at numerous Australian bird club and society gatherings and conferences; as the subject of the educational release *Learning About Cockatiels* by Mackay Productions; as past patron of the Native Cockatiel Society of Australia; as Secretary and regular presenter for the Aviary Bird Association of Australia Inc (1992–1997); and as a contributor to **Australian BirdKeeper Magazine**.

Diana has extensive knowledge of genetics, breeding programs and handrearing of many cockatoo species, including the Cockatiel, the Galah, Black, Sulphur-crested and Major Mitchell's Cockatoos, as well as rosella species, Sun Conures and Eclectus Parrots. This knowledge has also enabled Diana and her husband, Kim, to develop a range of incubator and brooder products under the name Kimani.

Her vast experience and dedication to responsible species management and conservation sees Diana active in the CALM breeding programs of the Long-billed Corella endemic to the Lake Muir region of Western Australia, the Red-tailed *Calyptorhynchus banksii naso*

Black Cockatoo and the White-tailed *C. latirostris* (Carnaby's) Black Cockatoo.

ABK Publications and indeed pet and aviary keepers of the delightful Cockatiel are indebted to Diana for her extensive knowledge and experience in the keeping and breeding of this species.

Contributing Authors

Dr Bob Doneley BVSc FACVSc (Avian Health)

Avian veterinarian, Dr Bob Doneley prepared the *Health and Disease* chapter and the *Wing Clipping* section. He also made invaluable suggestions and comments on the general content of this title.

Bob graduated from the University of Queensland in 1982 and worked in small animal practices in Bundaberg, Brisbane, Toowoomba and the UK, before opening his own practice in Toowoomba in 1988. He achieved Membership of the Australian College of Veterinary Scientists in 1991 and was awarded his Fellowship in 2003 after a four-year training program that took him to the UK once and the USA twice.

Also in 2003 he was appointed Adjunct Associate Professor in Avian and Exotic Pet Medicine at the University of Queensland Veterinary Teaching Hospital, where he conducts an avian and exotic pet clinic. In 2004 he was awarded the College Prize by the Australian College, for outstanding contributions to veterinary science in Australia. He has twice received Outstanding Service and Commitment Awards from the Association of Avian Veterinarians (AAV).

Bob is a member of the Australian Veterinary Association (AVA), the Australian Small Animal Veterinary Association, the Association of Avian Veterinarians Australian Committee (AAVAC), the Association of Reptile and Amphibian Veterinarians and the Association of Exotic Mammal Veterinarians. He is a past president of the AAVAC and the Darlings Downs branch of the AVA.

He has published numerous papers in veterinary journals, a veterinary textbook on bird medicine, and has written two chapters for another textbook published in the USA in 2006. He has spoken every year for the past 15 years at the annual conference of the AAVAC, and has lectured in the USA twice—at the International Conference on Exotics in 2002 and at the North American Veterinary Conference in 2006, both in Florida. Bob has also lectured at numerous veterinary conferences, avicultural gatherings and conventions throughout Australia.

His main veterinary interests include avian medicine, surgery and behaviour, and all aspects of exotic pet medicine, including reptiles, rabbits, ferrets and rodents. Readers of **Australian BirdKeeper Magazine** are familiar with the regular column that Bob prepares on all aspects of avian health. He has also advised and contributed to various **ABK Publications** titles over the years.

Dr Milton Lewis BSc (Hons) PhD

Dr Milton Lewis contributed the section *The Cockatiel in the Wild*, using information gleaned from his academic research and hands-on avicultural experiences.

Milton has had a long association with aviculture both as a keen exhibitor of canaries and finches and in his pursuits as a research scientist. At an early age while living in Canberra in the ACT, Milton kept a wide variety of finches and parrots. He enjoyed the challenge of keeping and breeding difficult species but along the way started with the more common species such as Zebra Finches and Cockatiels.

After completing an honours degree in Zoology at the Australian National University he studied the ecology of the Superb Fairy-wren for several years before accepting a PhD scholarship and a position as associate lecturer at James Cook University in North Queensland. This was a wonderful chance to increase both his scientific knowledge and research skills while at the same time moving to a climate where he could keep Gouldian Finches with relative ease.

While completing his investigation of the ecology and mating system of the Golden-headed Cisticola, a small Australian warbler, he undertook investigations of mate choice in Gouldian Finches. Following this successful investigation of the role of face colour in mate selection he was rewarded with the opportunity to apply his extensive research skills to investigating the ecology of wild Gouldians in the Northern Territory. Since finishing his research with wild Gouldians he continues to investigate the decline of Australian finches in the wild and is now concentrating his efforts on the Black-throated Grassfinch. He has also embarked upon unravelling the genetic determinations of plumage colour in Zebra Finches. Milton presents a regular column, *The Wise Owl,* published in **Australian BirdKeeper Magazine**.

Jim McKendry BTeach BAppSc

Jim McKendry read and commented on the *Companion Birds* chapter and prepared the section *Avoiding Behavioural Problems*. He operates Parrot Behaviour and Enrichment Consultations and a parrot behaviour and education web site located at www.pbec.com.au. His focus is on developing an understanding of the behavioural dynamics of parrots both in the wild and in captivity. As a parrot behaviour and enrichment consultant, he has provided consultative services to hundreds of parrot owners internationally, as well as contributing articles for numerous pet and avicultural publications, including a regular column, *Pet Parrot Pointers,* published in **Australian BirdKeeper Magazine**. Jim is also an active member of the International Association of Avian Trainers and Educators.

ACKNOWLEDGEMENTS

ABK Publications and the authors appreciate and thank the following people for their various contributions and support in the preparation of this comprehensive and informative title, in which research, expertise and experience involving the Cockatiel have been important factors.

Peggy Cross was co-author of the first edition of **A Guide to Cockatiels and their Mutations—their Management, Care and Breeding**. Her valued contribution has been widely appreciated, particularly by the breeders of mutation Cockatiels.

Thanks also to Mike Anderson, Laurie Bethea, George and Maureen Blair, Dr Mike Cannon, Barbara Carl, Terry Casey, Frank Garbers, Dr Greg Harrison, Brian Higginbotham, Hank Jonker, Dr Debra McDonald, Wayne Miller, Greg Paull, Leanne Phythian, Warren Power, Tom Roudybush, Jo-Anne Watts and Neil Whillans.

We are also most grateful for the photographic contributions from Tricia Belcher, Dr Mike Cannon, Peggy Cross, Richard Cusick, Phil Digney, Thierry Duliere, Donna Fowler, Janice Godwin, Valerie Grayston, Brian Higginbotham, Lawrence Jackson, Nancy Johnson-Mello, Stefan Kunz, Andreas Lindner, Otto Lutz, Dr Debra McDonald, Jim McKendry, Wayne Miller, Alfred Müller, Peter Odekerken, Conor Pearson, Pet City Mt Gravatt, Pet Directory, Leanne Phythian, Glenn Roman, Gerhard Rübesam, Gail Sibley, Sandra Trottier, Pieter van den Hooven, Dr Colin Walker, Jo-Anne Watts and Günter Wulf.

INTRODUCTION

The scientific name for the Cockatiel is *Nymphicus hollandicus* (Kerr 1792), although it was formerly known as *Leptolophus hollandicus*. The Cockatiel is now widely accepted as being the smallest member of the Cacatuidae family (cockatoos). Features that this bird shares with its larger counterparts are a movable crest, the presence of powder down in the plumage and the fact that both the cock and hen share incubation duties.

The Cockatiel is one of the most popular pet and aviary birds in the world. It is the perfect bird for a captive environment, suiting all types of interests and needs. Its charming personality, eagerness to breed and hardy constitution keep this species in constant demand.

Companion Cockatiels can learn to mimic human sounds, whistle operatic arias with enthusiasm, sound like a telephone (and also call their human friend to answer it), bark like the resident dog and mimic the calls of birds that are regularly in the vicinity. If housed indoors they will quickly learn to imitate the microwave beep or any other sound heard repeatedly in your home.

Normal Cockatiel cock.

They will cheerfully include you in their family and allow you to participate in the planning and raising of their young, sometimes to your embarrassment. If you need a loyal friend or a constant companion, a Cockatiel can make you the centre of its life. This is quite a responsibility, and before adopting a young bird as a pet, you should consider carefully whether you can live up to your pet's needs.

For those beginning in aviculture and wanting to enjoy a hardy, undemanding bird in a backyard aviary, the Cockatiel is an ideal subject. Provision of well-planned suitable housing, a nutritionally balanced diet and a variety of enrichment items will ensure that your Cockatiels live long and productive lives.

Breeders who want to learn, develop or improve a strain of birds will find the Cockatiel an ideal bird. With a long list of already recognised mutations and more being added, a lifetime could be spent developing this facet of Cockatiel husbandry alone.

Other Names

The name 'Cockatiel' is generally accepted to be an adaptation of a Dutch/Portuguese word that, translated, means 'little cockatoo'. Cockatiels are still known by regional colloquial names throughout Australia. Western Australians may refer to them as the Weero or Weiro, names reportedly adapted from aboriginal words. Eastern Australian states commonly call them Quarrions.

The Naturalist's Library (Jardine 1836) features a handpainted illustration by Edward Lear of the Red-cheeked Nymphicus. This bird was identified as *Nymphicus novae hollandiae*. Scrutiny of the illustration clearly shows that the bird is the Australian native Cockatiel. In other historical literature the Cockatiel has been known as Corella, Crested Parakeet, Crested Ground Parakeet, Grey Parrot, Yellow Top-knotted Parrot and Cockatoo Parrot.

Top: Mature cock showing no underwing markings.
Centre: Mature hen showing underwing markings.
Bottom: Young cock showing underwing markings starting to disappear.

Description
Length: 30–33cm
Weight: 89–100 grams

Wildtype Cockatiels are predominantly grey, with a pattern of white on the outer and secondary flight feathers. This pattern fits together to form a white wing bar that is clearly visible whether the bird is at rest or in flight. In birds with a heavy yellow suffusion, this wing bar is streaked with yellow.

Cock
The mature cock develops a bright yellow face and throat and an orange cheekpatch. The yellow and orange colouration in each bird varies in the depth and distribution of colour. The undertail feathers are dark grey. The upswept, movable crest is coloured bright yellow from the face to about half its length, where it shades into grey tips. The expressive crest is used for communication. The mandible and cere are dark grey, the iris dark brown and the feet and toenails dark brown.

Above and left:
Normal Cockatiel cock.

Hen

The wildtype hen is similar to the cock with a lesser amount and duller yellow colouration around the eye. The cheekpatch is also a duller shade of orange with a grey suffusion on the face. The lower back, rump and central tail feathers are pale grey, barred with white and pale yellow. The lateral tail feathers are yellow barred with grey. The undertail feathers are barred with yellow tones. The chest is shaded light grey and often speckled. The crest is primarily grey tinged with pale yellow and is as long and expressive as that of the cock. The hen retains the rows of underwing spots that are present in all wildtype juveniles.

Left: Normal Cockatiel hen.
Below: Juvenile Normal Cockatiel cock.

Immature

When newly hatched, wildtype Cockatiels have yellow down and dark eyes. Juveniles have a pinkish cere which can 'blush' when they are startled or injured. With colouration similar to the hen, young Cockatiels appear darker in colour until powder down is produced in enough quantities that it can be dispersed throughout the feathers. This distribution softens the Cockatiel colours and also serves to make the birds waterproof.

At about four months of age yellow streaking begins to appear on the face and the barred tail feathers are replaced with solid adult hen colouration. The mandible is horn-coloured tinged with grey.

Voice

Cockatiels can emit a relatively loud single call when alarmed or greeting another bird. However, for the most part, their call is fairly quiet and inoffensive when compared to other members of the cockatoo family. Cocks are capable of developing a much more elaborate repetitive whistle as part of their courtship display.

COCKATIELS IN THE WILD

Origins

Before the advent of cellular/DNA evidence used by modern taxonomists, the only data used to better understand the familial relationships between groups of organisms was the use of morphological features and, in some instances, behaviours. This had been the case for the Cockatiel where controversy existed, the major question being whether it was a cockatoo or a parrot!

Cockatoos and Cockatiels share several characters such as an erectile crest, a gall bladder and a non-pericyclic iris that help distinguish the group from other parrots. However, these similarities have not always been obvious and some authors have attempted to classify the Cockatiel into a lineage with other parrots. Features such as the ear passage to and from the eardrum (auditory meatus), the structure of the pineal gland located at the base of the brain that produces melanotonin, the powder down in the plumage, the sequence of feather loss during the moult and the possession of wing spots in hens and juveniles have all been used in such classifications to suggest that the Cockatiel was not a cockatoo (Adams *et al* 1984).

At present the Cockatiel is recognised as a cockatoo, belonging to the family Cacatuidae, which contains 18 species, all with an Australasian distribution (Brown & Toft 1999). Adams *et al* (1984) suggested a lineage containing two major groups, which in its simplest terms is composed of *black* (Calyptorhynchini) and *white* (Cacatuini) cockatoos; the Cockatiel is located within the Calyptorhynchini. The first group was formed by the genera *Probosciger* and *Calyptorhynchus*, while the second group contained *Cacatua*. Modern allozyme evidence (Adams *et al* 1984), mitochondrial DNA (Ovenden *et al* 1987) and a tandem repeat in parrot nuclear DNA (Madsen *et al* 1992; Dixon 1994) also support the arguments that the Cockatiel is a *small cockatoo*.

Once the audience was satisfied that the Cockatiel was a small cockatoo, the real question of the species' relationship to other members of the family could be addressed. Smith (1975) separated members of the family on the loss of 'Dyck texture' (feather structure which produces blue and green colours in other parrots), acquisition of male incubation, development of the behaviour for indirect head-scratching and the loss of tail barring. In this classification indirect head-scratching was the character used to produce the branch containing the Cockatiel and thus separate this species from all other cockatoos. Using differing methodologies, previous authors have all come to a consensus that Cockatiels are closely related to *Calyptorhynchus* in the branching sequence but still remain as a monotypic genus.

In the most recent review of the cockatoo classification, similar results to previous authors were recorded and perhaps have resolved some of the underlying criticisms of prior research. Brown and Toft (1999) concluded that the Palm Cockatoo *Probosciger aterrimus* is the first branching sister group in the phylogeny of the cockatoos and therefore perhaps the species most closely resembling the early cockatoo form. The next branch consists of a subclade (subgroup) containing the Gang Gang Cockatoo *Callocephalon fimbriatum*, the Red-tailed Black Cockatoo *Calyptorhynchus banksii*, the Major Mitchell's Cockatoo *Cacatua leadbeateri*, the Galah *Cacatua roseicapilla* and the Cockatiel. The other black cockatoos are included in this subclade but were not discussed in the results because the authors were unable to source material for all the species and instead chose the option of using a single representative from an obviously closely related group. Two other later subclades in the topology include the remainder of the white cockatoos.

Biogeographically, the origin of the cockatoo family appears to have formed within Australia and radiated to islands within Asia and the Pacific region. Historically, Australia has undergone several very dry periods where the number of bird species has significantly declined. The species surviving these dry periods did so by remaining within islands of vegetation scattered throughout the continent that were shielded from the full impact of the weather. Areas such as the Flinders and MacDonnell Ranges or sections of the Great Dividing Range formed species refuges. Sometimes the periods of isolation from sister species appear to have been long enough for new species to evolve to a point where once the climate improved and these species came back into contact with each other, interbreeding was not possible.

This may have been the scenario for the cockatoo family and the Cockatiel was very likely one of the species using inland refuges. The more recent expansion of the family by the subclade for white cockatoos occurred during a period when land connections between Australia and New Guinea existed and then further colonisation occurred throughout the adjoining archipelagos. The earliest known fossil record—from Queensland—is of a form similar to *Cacatua* from the early to middle Miocene Epoch (Boles 1993). However, the phylogenetic evidence of Brown and Toft (1999) and further fossil evidence (Boles 1993) suggest earlier origins for cockatoos from perhaps the middle Tertiary Period.

Present Distribution

The Cockatiel is found throughout much of inland Australia, although in extremely dry regions such as the Gibson Desert in Western Australia this species is less likely to be encountered. Other regions of lower abundance are along the coastal fringes, especially in the south-east of New South Wales and Victoria, Cape York, Arnhem Land in the Northern Territory and the south-western corner of

Western Australia. Seasonal movements for this species have been noted by several authors, which are reflected in greater detail by data for the Australian Ornithologists Union (Barrett *et al* 2003). This data suggests a generalised concentration of Cockatiels within the western edge of the Great Dividing Range and northern Victoria during the summer breeding period. As autumn approaches (March–May) more birds are seen to the north-west of New South Wales and reaching into Queensland and the Northern Territory. During winter (June–August) the Cockatiel can be seen throughout most of the states and territories although there are fewer sightings in the southern half of Western Australia. During this period small numbers of Cockatiels reach the furthest points in their distribution, even as far north as Darwin. Around the Katherine area of the Northern Territory small groups and pairs of Cockatiels are regularly observed at waterholes early in the morning during August and September but are absent for much of the rest of the year. Similar patterns of occurrence are observed in Townsville. As spring approaches (September–November) birds begin to move towards the east again. Although the Cockatiel does not occur naturally in Tasmania, there have been occasional records of sightings in that state. These are presumed to be aviary escapees (Forshaw 1981).

Food and Feeding Habits

In the most detailed account of the feeding ecology of the Cockatiel, Jones (1987) found that this species had two main periods during the day in which they feed. The morning feast commences about 30–50 minutes after sunrise and may last for about 1.5 hours. During this period birds consumed about 2.72 grams of dry seed. An afternoon feeding session commences about 60–90 minutes before sunset and continues for about the same time period as the morning session. During the afternoon feeding, the birds consumed about 4.25 grams of dry seed, about double that eaten in the morning. On some occasions birds were observed feeding for short, 10-minute periods during the middle of the day and just before sunset. At the conclusion of the final short feeding period birds return to their roosts where they remain for the night.

Birds studied within the Moree region of north-western New South Wales displayed a strong preference for feeding on sorghum, which is commercially grown in that district. In this area sorghum was preferred before the seed was completely hard and ripe. This soft seed formed about 60% of the diet throughout most of the year, with commercial grain such as sunflower and millet comprising about 80% of the diet. In total, 29 different types of seed were found in the diet of Cockatiels in this study but other seeds have been observed being eaten as food in other areas of their distribution (see Table 1). The three most important species of seed, which helped account for the remaining 20% of the diet, were Queensland Bluegrass, Native Millet and Hood Canary Grass. In the Northern Territory this species has been observed feeding on Spear Grass at the end of the Dry Season. The crop contents of birds collected by Jones (1987) also revealed the presence of pieces of charcoal and quartz that may aid digestion of hard seeds.

Jones (1987) noted that all seed from native grasses, wheat and oats was taken from stocks off the ground while sorghum and sunflower were preferentially removed from seed heads that were still growing. Flocks were generally composed of about 10–27 birds, but in periods of seed shortage the size of flocks dramatically increased to as many as 500. In some instances Cockatiels have also been observed foraging with other parrot species such as Red-rumped Parrots *Psephotus haematonotus* (Forshaw 1981) or in mixed flocks of finches and Hooded Parrots *Psephotus dissimilis* in the Northern Territory (Lewis pers. observ.). When feeding on the ground, flocks move forward in a slow methodical fashion as each bird locates a seed and dehusks the kernel before it is eaten. If seed is to be removed from a stem the task is completed by chewing through the base of the stem and then removing the seed from the felled head. Sunflower seed is removed from the flower head while the bird is perched on the upper rim of the capitulum (seed head). Only seed from the upper sections of the capitulum is eaten and birds were never observed hanging upside down removing seed from the lower part of the flower.

Table 1: Foods Eaten by Wild Cockatiels

Seed Species	Method	Reference
Acacia sp.	Unknown	Juniper & Parr (1998); Forshaw (1978)
Amaranthus	Crop contents	Jones (1987)
Animated Oat	Observations	Jones (1987)
Annual Spear Grass	Observations	Lewis (unpublished data)
Aristida sp.	Crop contents	Forshaw (1981)
Awned grasses (*Astrebla sp.*)	Crop contents	Jones (1987)
Barley Grass	Observations	Jones (1987)
Browntop	Crop contents	Jones (1987)
Button Grass	Crop contents	Jones (1987)
Clover	Crop contents	Jones (1987)
Common Burr Medic	Observations	Jones (1987)
Cotton Panic	Observations	Jones (1987)
Curly Windmill Grass	Observations	Jones (1987)
Dock	Crop contents	Jones (1987)
Foxtail	Crop contents	Jones (1987)
Ground Cherry	Observations	Jones (1987)
Hood Canary Grass	Crop contents	Jones (1987)
Hoop Mitchell Grass	Crop contents	Jones (1987); Smith (1978)
Lindley's Saltbush	Crop contents	Jones (1987)
Mistletoe	Observations	Forshaw (1978)
Native Millet	Crop contents	Jones (1987)
Oats	Crop contents and observations	Jones (1987)
Perennial Cupgrass	Observations	Jones (1987)
Queensland Bluegrass	Crop contents	Jones (1987)
Rat-tail Grass	Crop contents	Jones (1987)
Small Sago Weed	Observations	Jones (1987)
Sorghum	Crop contents and observations	Jones (1987)
Sunflower	Crop contents and observations	Jones (1987)
Tar-vine	Crop contents	Jones (1987)
Trirophis sp.	Crop contents	Jones (1987)
Wallaby Grass	Observations	Jones (1987); Forshaw (1981)
Wheat	Crop contents and observations	Jones (1987); Juniper & Parr (1998)

Breeding

Cockatiels breed as pairs when climatic conditions are suitable. Many authors suggest that breeding is perhaps triggered by rainfall events (Simpson & Day 1986; Forshaw 1978; Smith 1978), which is not unlikely if compared to the breeding patterns of other Australian inland bird species. However, even if there are breeding events that are unseasonal there is also an annual pattern of nesting indicated by observational data (Barrett et al 2003). Where breeding is cyclical it corresponds to warmer austral months between August and January (Barrett et al 2003; Forshaw 1978, 1981). In south-eastern states where the majority of breeding records occur, this also corresponds to periods of spring rain, increased grass seeding and lower night temperatures.

Cockatiels nest in hollows usually within *Eucalyptus* trees preferentially close to water (Forshaw 1981). The position or size of the nest hole entrance does not appear to be significant as I have observed nesting pairs of Cockatiels using a wide variety of hollows. I have also observed Cockatiels using hollows located from 2–15 metres above the ground. Clutches contain 4–7 eggs laid on decaying wood inside the hollow and hatch within 18–20 days after incubation has commenced. Hens brood the eggs during the night and cocks brood during the day. Young leave the nest 4–5 weeks after hatching and remain dependent for several more weeks although data for this period is very limited (Forshaw 1978, 1981). The adults feed their young on a diet consisting of both native and introduced seeds.

The pair bond appears to be relatively strong with pairs observed together throughout the year (Smith 1978) but this has not been rigorously tested.

Wild Cockatiels at Broken Hill.

Status

The Cockatiel is a common Australian cockatoo, occurring throughout inland parts of the continent. Increased agricultural production of grain has possibly influenced the present distribution and abundance to a point where this may be one of the most common wild parrots in Australia. The high population numbers indicate that this species is safe within the wild. However, because of the large numbers in concentrated agricultural regions there has been some concern that the Cockatiel may be a pest in areas of high grain production.

COCKATIELS IN CAPTIVITY

Normal Cockatiel hen (left) and cock.

AVIARY BIRDS

History

Cockatiels have been described in literature since the 1700s. They were displayed at the *Jardin des Plantes* in Paris in 1846, where they were bred successfully. The Cockatiel was considered an established aviary bird in Europe by 1884 and was established in the USA by 1910. Cockatiels have been aviary bred in Australia since 1901. Interest in this underrated bird accelerated when the Lutino mutation appeared in 1958.

Show Standard

The Cockatiel now comes in many variations of colour and pattern. However, the wildtype bird should be used as a baseline for size and stature. Domestic diets and controlled breeding have provided the atmosphere for the Cockatiel to increase in size, a development no doubt encouraged by market demands, but one which needs to be closely monitored to retain the integrity of the wildtype bird. Standards should be documented in an attempt to keep the perfect and ideal Cockatiel just that—not bigger or longer.

Every effort should be made to ensure that our perfect Normal Cockatiel does not succumb to a fate similar to that of the now, almost unrecognisable Show Budgerigar. The Cockatiel is a small to medium-sized parrot and should measure 30–33cm in length. Some show standards encourage a larger bird, up to 35–38cm. These measurements include the tapered, movable crest of 6cm or more and the long and slender tail which

The Normal (wildtype) bird should be used as a baseline for show standard.

should measure approximately 17.5cm and account for half the entire length of the bird. The range of weights in captive birds is 80–120 grams. The head should be large and rounded with no flat or bald spot on the top or on the back of the skull.

The opening description from the Australian National Cockatiel Society Show Standard is succinct and ideal: 'The Cockatiel is a sleek bird with straight back and full chest, giving an overall look of a strong bird able to cover vast areas in its daily search for food.'

For those people interested in

Winning Champion Bird (2006) bred by Jo-Anne Watts of Brisbane.

showing their birds competitively, a Cockatiel Society in your area can supply you with a copy of the National Show Standard. For those interested in simply breeding good quality birds, the show standard can be used as a guide. In time you will get a feel for quality birds and automatically know what to look for. The most common mistake made is to breed for colour, markings and size with less emphasis on good conformation.

Behaviour

One of the Cockatiel's best qualities is its non-aggressive nature, making it an ideal specimen for a colony, small group or mixed aviary—provided that this does not place them in danger

Above: Stretching is done in conjunction with preening.
Left: Yellowcheek pair. Mutual preening is a favourite pastime.

from other occupants. Mutual preening is a favourite pastime and any feather loss that becomes apparent is normally the result of excessive preening rather than aggression.

Although it is not unusual for Cockatiels to sleep for periods during the day, they are active birds that should be provided with environmental enrichment, eg branches with buds and seed pods to chew on. This will occupy their attention for long periods and thus reduce the likelihood of preening feather damage or ingestion of zinc from wire chewing. Although they are not strong enough to significantly damage wire that is 1.3mm thick or more, bored Cockatiels will sometimes begin to lick the wire obsessively and can ingest small lumps of welding slag, resulting in heavy metal toxicity.

In colony situations, insufficient nesting and roosting sites can cause squabbling. While this rarely results in any physical injury, the stress that it causes should be avoided.

Acquiring Stock

Walk into most pet shops anywhere in the world and you will find Cockatiels offered for sale. These are usually Normal (wildtype), a few of the more common mutations such as Lutino, Pearl and Pied and occasionally some of the less common mutations.

For someone wanting a few cheerful and attractive birds for a backyard aviary, purchasing from a dealer or pet shop will probably suffice for their needs. However, for those interested in the breeding of colour mutations, it is best to acquire your birds from private breeders who close band their birds, keep accurate records and can guarantee the genetic background of their birds.

For those wanting a pet, or a pair of pets, the best birds to acquire are undoubtedly handreared chicks. These are already accustomed to humans and used to being handled.

The more gregarious individuals are immediately apparent, as they usually cling to the wire demanding human attention long after they have been weaned.

Birds that have been raised and weaned by their parents and tame, handraised birds are both suitable for an aviary environment.

Regardless of the reason for acquiring the bird, there are a few basic features to look for regarding the health and wellbeing of the individuals that you are selecting. One of the Cockatiel's most endearing qualities is an active, happy disposition. When not sleeping, which it does through the middle of the day and at night, it should be actively moving about, preening, feeding and investigating its area, making it relatively easy for the novice to select a healthy bird.

If you are choosing birds in summer, do so early in the morning before the heat has set in. At this time of day the

Platinum Cockatiel hen.

birds should be active and an individual that appears sleepy and unwell will be more obvious. Catching and transporting birds on hot days is stressful and unwise. In very hot conditions, most Cockatiels will rest on both feet with their wings held out, their feathers held close to their body and their beaks slightly open. This is definitely not their most flattering stance. Remember that even if the birds are not showing signs of heat stress they can rapidly become overheated when being netted and even handreared birds will

*Above: Paleface Grey Cockatiel hen.
Right: Recessive Pied Cockatiel hen.*

become stressed while being transported to a new home.

In winter it is better to view birds late in the day when they begin their late afternoon feed. Again, at this time they should be active and not feeling the cold. A bird that is looking fluffed and dozing on two feet should be avoided, as this is likely to be a sign of ill health.

The bird that you choose should be a good size and in good feather relative to its age. Cockatiels increase in size for at least 12 months. Even youngsters that have had severe health or injury setbacks will often reach normal size in this period if fed a good diet. Therefore, look for a bird that is a reasonable size for its age and place much more emphasis on its conformation, which will change to a lesser degree as the bird matures.

Young birds can look terribly small and scruffy when they go through their juvenile moult between 4–6 months of age. If you know for a fact that a bird you are hoping to purchase is of this age then you can possibly excuse its appearance. However, it is probably best to choose only birds that are in good feather. Adult Cockatiels moult in January–February (summer in the Southern Hemisphere), although unusual weather patterns can often bring about early or late moults. Choosing a bird in perfect feather at this time of year can be difficult.

Select birds in good feather such as this Whiteface Grey Cockatiel cock.

Extensive areas of missing feathers, particularly around the back of the neck (with the exception of the Lutino mutation), are often the result of overpreening by other birds, particularly when the birds are being kept in overcrowded conditions with little to keep them occupied. This in itself, although unsightly, may be harmless, although the perpetrator of the plucking will often turn into a plucking parent. On the other hand, any evidence of selfmutilation or abnormal feathering is likely to be a health problem and should therefore be avoided. Avoid birds with eye infections, a nasal discharge or missing feathers around the eyes. These symptoms may indicate a serious health problem, including Psittacosis, a highly contagious disease in parrots. Also to be avoided are birds showing signs of heavy or laboured breathing—a further symptom of a health problem. (See *Health and Disease* on page 178.) Observe the birds closely for lameness and toe and beak abnormalities. The vent and surrounding feathers should be free from faecal pasting. When you physically examine a bird that you have chosen, you should feel a layer of firm muscle and flesh on either side of the keel bone. Sharp protruding keel bones are a sign of ill health or a serious worm problem or both.

When purchasing birds sight unseen from breeders too distant to visit personally, you can only rely on the honesty of the breeder concerned when verbally representing their birds. All birds emerge from transport cages looking ruffled and stressed. Unless the birds are obviously in poor condition or not at all what you wanted, final judgement on your new purchases should be reserved for at least a few days, preferably weeks, to allow the birds to settle in and groom themselves. However, under no circumstances should you accept birds that you feel were misrepresented.

It is highly recommended that your newly acquired birds be examined by an avian veterinarian. (See *Health and Disease* on page 178.)

Quarantine

All new stock, regardless of species, should be quarantined for a period of 6–8 weeks, before being introduced to your other birds. Quarantine serves two purposes. Firstly, it minimises the risks of introducing disease and parasites into a collection. Secondly, it gives a new bird a chance to become acclimatised to a new management and dietary regime, before having to deal with the stress of establishing a social pecking order in its new home. Both of these purposes are equally important but are often overlooked by inexperienced aviculturists in their haste to see a new bird out in an aviary.

A suitable carry or freight box.

Quarantine cages can be as complex as a separate flight, or as simple as a large cage. Whatever you use, it must be geographically isolated from your other birds. Placing a new bird into a spare flight in a bank of aviaries does *not* constitute quarantine. Birds in quarantine cages must be managed in exactly the same way as your other birds—same kind of feed dishes, same feeding times and diet—but must be fed and watered last. No matter how keen you are each morning to see your new purchase, it must be the last bird attended to. This simple step prevents you backtracking to your aviaries, perhaps with contaminated faecal material on shoes, clothing or hands.

A suitable quarantine cage.

The quarantine period provides a good opportunity to treat your bird for parasites. Spraying the bird with a good quality lice and mite spray every two weeks for three treatments will prevent external parasites being introduced into your collection. The bird should be wormed at least twice, and its efficacy then evaluated by a faecal test by your local avian veterinarian.

If you introduce any other new birds into quarantine at this time, they should be placed in their own quarantine cage in a separate location. If this is not feasible,

A plastic crate is easily converted to a carry or quarantine cage with three closed sides providing security.

the new arrivals must be placed in the same cage as the previously quarantined bird. You must therefore re-start your quarantine clock, ie the six weeks start from the time the new birds arrive.

Performed carefully, thoroughly and with some forethought, quarantine can be the simplest and cheapest form of disease control that an aviculturist can implement.

HOUSING

One of the most common words used to describe the Cockatiel is 'hardy'. This term often translates to providing poor housing conditions, in the belief that Cockatiels are undemanding birds that will maintain good health and reproduce happily in anything from a wired-in fruit box in the backyard shed to a nestbox in the corner of someone's lounge room!

These suspended and conventional aviaries are used to house young unpaired and bonded breeding birds respectively.

While it is true that these tough little birds will endeavour to raise their young under all kinds of adverse conditions, supplying them with adequate housing is the minimum responsibility of all bird keepers. Supplying them with ideal housing will result in happier, healthier birds that, in turn, will reward their keepers with greater productivity.

There are three main methods of housing Cockatiels—conventional aviaries, suspended aviaries and pet cages. There are several essential requirements that are common to all three.

These are:
- Protection from inclement weather, excessive heat and draughts;
- Protection from vermin and predators;
- Vitamin D_3 in the form of access to sunlight particularly early morning sun or a vitamin and mineral supplement;
- Escape-proof safety areas;
- Easy access to fresh food and water;
- Adequate exercise area; and
- Privacy.

Conventional Aviaries

Probably the most common form of housing for Cockatiels is the conventional aviary, either a single backyard aviary housing a pair, a small colony or a mixed collection or a bank of breeding aviaries.

Aviaries housing a breeding pair should be at least 1.2–1.5 metres wide, a comfortable height for the keeper to walk in and 2.4 metres long, preferably longer, as Cockatiels will often revert to climbing rather than flying from one end of the aviary to the other if the flight is too short. Lack of adequate exercise will result in overweight birds that can develop breeding problems. However, aviaries that are too long can occasionally result in fatalities in young birds who gain too much speed without enough control when leaving the nest.

Birds that are not paired and set up for breeding are frequently housed as groups in large aviaries with long flight areas to encourage exercise. Because Cockatiels are a flock species, this size aviary can also be used for housing a group of young or non-breeding birds (6–10 individuals). Cockatiels are generally not aggressive although the more birds you place into a single aviary, the more stressful the environment can be. The birds should be monitored closely for signs of bullying and resulting stress.

Fully covered flights are recommended for conventional aviaries. A roof will protect the birds from predators such as raptors, owls and cats and prevent droppings from wild birds contaminating the aviaries. One disadvantage of this set-up, however, is a lack of access to sunshine throughout the day. It is beneficial for sunlight to enter part of the aviary for at least a short time each day. This is possible, even if fully roofed aviaries are used, by carefully considering the location of the aviary within your yard to allow maximum access, preferably, to early morning sunlight. If angled sunlight enters the aviary at a certain time each day, a perch should be provided in the sunlit area to allow the birds to sun themselves.

In many countries, because of extreme cold, Cockatiels are successfully bred indoors with no access to sunlight at all. In this case it is important to provide UVB lighting and a dietary supplement of vitamin D_3 to prevent calcium deficiency unless the birds are being fed on a pelleted diet which already includes adequate vitamin D_3 supplementation.

Cockatiels are very fond of bathing, and in a fully roofed aviary the birds are not able to bathe in the rain. For this reason, a fine sprinkler should be positioned so that the spray enters an area within the aviary to allow the birds to bathe in warm weather. This is an enriching activity for the birds and helps to maintain their plumage in good condition.

The addition of a dim night-light will help eliminate deaths, injuries and abandoned young due to night frights caused by predators and vermin.

Suspended Aviaries

Suspended aviaries are used extensively by many as they are well suited to breeding Cockatiels. One of the major criticisms of suspended cages or aviaries is that if they are too small they can cause feather disorders and overweight birds. The simple solution to this criticism is to make them the same length as conventional aviaries. However, they can be narrower than conventional aviaries without disadvantaging the birds' flying space as there is not the need to allow room for the birds to fly past a person standing in the aviary. This allows for a greater number of aviaries to be housed in a smaller area. Suspended aviaries for an individual pair or a small group should measure at least 2 metres long x 90cm wide x 1.2 metres high. Cockatiels can be colony bred, but for accurate and controlled breeding results, one pair per aviary is essential.

An advantage of suspended aviaries is that the birds appear to be secure and very

Suspended aviaries in an attractive garden setting at Maureen and George Blair's premises.

settled. This is perhaps because their environment is not constantly being invaded by a large foreign body, often armed with a rake or a broom. Probably the most intrusive item that enters suspended aviaries is an occasional catching net. Even very nervous birds will perch close to the front of the aviary, making close observation of the birds easy.

This suspended aviary block is divided into six flights.

Although suspended aviaries can be built as multiple divided units, this makes internal cleaning a little difficult. It is recommended that the cages be built as individual units or small sets of two or three cages so that they can be easily moved about or lifted down for cleaning. Ideally, frames are constructed from 25mm aluminium square tube with a large door fitted at one end. This enables access to the aviary to replace the perches and scrub the wire when necessary. An advantage of individual suspended aviaries is the provision of double wiring when placed in a set—this costs no more than the wiring needed for a row of conventional aviaries.

Because suspended aviaries are less likely to be contaminated by wild bird droppings—these fall through the wire to the ground below—only one-third of the roof needs to be covered. The birds then have direct access to full sunlight during the day and can bathe in the rain if they choose to. In this situation a double-wire roof over the open section is provided to protect the birds from predators. All-wire suspended aviaries provide no privacy, shelter or protection from inclement weather conditions or predators. This type of aviary should be partially housed inside another structure for the safety of the birds.

One disadvantage of suspended aviaries is that the birds are more difficult to catch. This can be overcome by positioning someone at the opposite end of the aviary to encourage the birds to move down to the end where the catcher is located.

Hygiene is one of the greatest advantages of suspended aviaries, with most droppings, seed husks and discarded food items falling through the wire. If larger pieces of uneaten food are removed daily and aviaries hosed down once a week, there is little risk of bacterial or fungal infections resulting from poor hygiene. The debilitating cycle of worm infestation will be broken as droppings will fall through the wire cage bottom. Suspended aviaries are also much easier to protect from mice and rats and it is easier to place baits in this type of aviary complex without any risk to the birds.

Suspended aviaries with gravel laid underneath enable easy maintenance.

Aviary Materials

Like all parrots and cockatoos, Cockatiels enjoy chewing wood. Therefore, the preferable construction material for the aviary framework is either steel or aluminium. The benefit of aluminium is that it does not rust but it is more expensive, not quite as strong as steel and its light weight makes the aviary more susceptible to being lifted and overturned in windy conditions. Steel, available in galvanised lengths that are less likely to rust, can also be powder-coated for a more attractive finish.

These lightweight suspended aviaries, constructed from aluminium tube with plastic joiners, can be easily transported or relocated. The base can be secured with tent pegs.

Corrugated Colorbond™ sheeting in a multitude of colours can provide an attractive finish to the sheltered areas.

Galvanised cage fronts can be fastened into the aviary wire structure to provide an economical access door to the feeding area.

Wire

Galvanised aviary weldmesh with a minimum diameter of 1.3mm can be fixed with hexagonal self-tapping screws to the inside of the aviary frame. The mesh aperture should be 12.5mm x 12.5mm and not greater than 25mm x 12mm, as Cockatiels can easily get their heads through a 25mm x 25mm aperture. Apart from the danger of being beheaded by predators, the birds have been known to hang themselves by putting their heads through one hole in the wire and back in through the adjoining hole. When they realise that they are caught they can panic and strangle themselves or break their necks by flapping in an effort to release themselves. Although not necessary for restraint purposes, 12.5mm x 12.5mm has the added benefit of restricting larger rodents and the talons of hawks and owls access to the aviary.

Aviary mesh can be painted black to increase visibility and make the aviary more attractive. However, care must be taken to ensure that the paint chosen is both non-toxic and suitable for metal surfaces. Foam paint rollers are the easiest tool to use to apply the paint evenly to the wire.

When choosing weldmesh for the aviary, be aware that new wire, when chewed, can poison your birds. Scrub the new wire with vinegar to neutralise the zinc oxide. This will assist in preventing your birds from succumbing to Heavy Metal Poisoning.

Shelter and Insulation

Both conventional and suspended aviaries must be designed to suit the climatic conditions of the region in which they are located. In most areas of the Southern Hemisphere aviaries should face north, if at all possible, in order to gain maximum access to winter sunlight.

In very hot climates, a double roof is recommended, especially if you want to encourage your birds to breed. A double roof reduces the heat radiation in the aviaries considerably, keeping the temperatures stable. The addition of some form of insulation can further enhance this effect. However, many insulation materials are attractive to rodents, providing them with an ideal breeding environment—and nesting material. A sprinkler system on the roof can also reduce the heat. However if the sprinkler is combined with only a slight breeze, the cooling effect is multiplied and even in summer conditions wind chill can affect young Cockatiels that do not yet have the waterproofing protection of their powder down. Some carefully positioned trees, including a few deciduous species, can provide the best solution—shade from the midday sun in summer and full sun in winter.

This fully roofed aviary provides protection from extreme heat and inclement weather.

Shelter should also be provided at the back and at least one-third of each side of the aviary. This provides the birds with additional protection from the weather. Cockatiels are far more susceptible to draughts than very cold conditions. Originally a desert bird, the Cockatiel tolerates heat and cold quite well. The sheltered area will also provide them with some security and privacy. Nestboxes hung in the sheltered area are usually accepted much more readily than those hung in the open section of the flight.

Walkways

Cockatiels are notorious for shooting out of the aviary over the shoulder of the keeper. A secure safety area for individual aviaries or a walkway for banks of breeding aviaries is highly recommended. It will help prevent the loss of a bird that has escaped through an open door of a flight. Many breeders also use the walkway to exercise their birds. The birds generally return to their own aviary when someone approaches and the door can be simply closed behind them. Tame birds can also be released into the area where they can interact with and be handled by the keeper.

The walkway will protect nestboxes and feeding stations from the weather. Wire feeding verandas can also be attached to the aviaries in the walkway, providing easy access to food and water dishes. This allows you to service the needs of your birds without entering the aviary, except for cleaning and nest inspection purposes. Breeding birds that are not tame become much more settled in their environment if they are not disturbed too much.

The walkway should be wide enough to store items and allow the passage of a feeding trolley or wheelbarrow.

To prevent birds from escaping, the access to the aviary is protected by a safety area.

Perches and Other Aviary Furnishings

Natural, non-toxic perches are recommended for both cages and aviaries. The varying dimensions of the branches should range upwards from 15mm in diameter to provide exercise for the birds' feet. Bark also provides enrichment for the birds in the form of hours of chewing entertainment. For this reason the perches should be replaced regularly.

Perches should be located in both the sheltered and open flight areas, at least 1.8 metres from ground level and positioned far enough apart to encourage flight. Positioning perches lengthwise in an aviary can be beneficial in that they require the bird to alter its flight pattern from flying forwards and backwards from perch to perch. They must change direction in order to land, which involves exercising different flight muscles. However, there is a downside—lazy birds will use the perches to negotiate the aviary without flying.

Because Cockatiels are avid chewers, plants are not likely to survive in the aviary. However, the birds should be provided with fresh branches complete with leaves, flowers and buds. Branches of most non-toxic trees, eg eucalyptus and bottlebrush, are suitable for this purpose. A regular supply of branches will not only enhance the birds' environment but will also keep them occupied for long periods and satisfy their chewing urges.

This natural branch perch is bolted to a sturdy perch holder.

This style of perch holder enables easy replacement of perches.

Branches can be placed in metal tubes attached to the side of the aviary or in a freestanding receptacle for holding fresh browse. This is made by setting metal tubes in a bucket filled with concrete.

Commercial hanging toys can also be included for interest. However, be aware of items that contain toxic metals and sharp objects and toys manufactured from fibrous material such as rope and twine. These items can become tangled in toes, legs and around the necks of birds once they start to fray. It is important to check the safety of any toys periodically.

Nestboxes

Cockatiels will readily accept a wooden nestbox. They are light and easy to mount, maintain and inspect. (Natural nesting logs are an unnecessary addition to the aviary, except from an aesthetic point of view.) Use nestboxes that are virtually identical so that if one needs to be replaced during the breeding season there is very little likelihood, if any, of the new nestbox being rejected.

Nestboxes should measure approximately 25cm square x 45cm high. This size of nestbox accommodates five or more large chicks who can generate considerable heat, particularly when feathering up. These dimensions give the chicks enough room to move apart, allowing some airflow between them.

Boxes can be constructed from untreated wood, eg chipboard, plywood and pine, and are hung vertically in the aviary. An entrance hole is cut near the top of the nestbox and a hardwood landing perch fixed below it. Two or three slats or steps are fixed internally to assist the parents to come and go without jumping on top of eggs. This also helps the youngsters practise climbing up and down before the big jump into the wide world.

Suitable Cockatiel nestbox design.

Wood is not recommended for the internal steps, as bored, incubating parents often chew the steps, leaving the nailheads exposed. A chick can impale itself on one of these exposed nailheads. Alternatively, screw an aluminium 'J' section onto the inside of the box. The screws face out of the box and the birds are able to climb down easily without the addition of any wire—no risk of toes and leg rings getting caught. If you wish to use timber it is worthwhile making the effort to find hardwood pieces for the landing perch and the steps leading into the nestbox.

In conventional aviaries the nestboxes should be mounted within the sheltered area. They can also be mounted outside the aviary in the service walkway with the entrance facing into the aviary. Provided that there is somewhere for the birds to land at the nest entrance, this type of nestbox is usually well accepted and allows the keeper to inspect the nestboxes without entering the aviary and disturbing the birds.

With suspended aviaries the nestboxes are almost always mounted on the outside, preferably in the sheltered area in the top left-hand corner. If you wait until the parents have left the nest, chicks and eggs can be inspected without causing any stress to the parents at all.

This nestbox design, made from plywood, provides a natural perch at the entrance hole and rear inspection access.

If nestboxes are mounted externally on suspended aviaries, the inspection hatch is at the back, close to the level of the chicks. If nestboxes are hung inside the aviary the inspection hatch needs to be on the front or the side. Include a lid that can easily be propped slightly open in excessive heat. When using this method be sure to check the young to confirm that they are being fed. Some parents refuse to enter their nest if the lid or inspection hatch is open or slightly ajar. (See also *Nestbox Preparation* on page 46.)

Substrate

Many bird keepers using conventional aviaries prefer earthen floors so that the birds can forage on the ground, as they do in the wild. However, for health reasons as well as rodent control, concrete floors are much safer. It is preferable to supply the birds with a good quality mineral grit that meets any digestive requirements. With the correct gradient on a concrete floor, droppings and other debris can be hosed away regularly, reducing the likelihood of worm infestation.

For those who prefer a more natural appearance to their aviary, there is an alternative. If there is a sheetmetal skirt around the base of the aviary, attractive, well-drained river pebbles can then be spread over the concrete floor to a depth of approximately 3cm. The metal skirt provides a base to retain the pebbles which can be hosed regularly. This prevents the potential build-up of parasite eggs which are difficult to remove from an earthen floor.

Rodent Control

Mice and rats are the bane of a bird keeper's life and once established are almost impossible to eradicate. Therefore, it is best to try and prevent them from entering the aviary in the first place. Feeding stations should be placed out of the reach of rodents, as it is the food that attracts them. This is only effective, of course, if you remove spilt food on a daily basis.

With an earthen floor, concrete footings extending 60cm into the ground may help prevent rodents from digging under the wire. Concrete floors are more effective. However, if the wire goes down to the ground, neither will have any impact on preventing mice from entering the aviary. Mice will easily pass through any form of wire other than 5mm square mesh and juvenile rats can pass through 25mm x 12.5mm square mesh—only 12.5mm x 12.5mm mesh will exclude them.

To rodent proof the aviary, a sheetmetal skirt at least 45cm high with a rodent guard at the top (a piece of metal flashing that protrudes at a 45° angle) should be placed around the base of the construction. Combined with concrete footings or a concrete floor, this is quite an effective method. However, the aviary must be positioned away from anything that the rodents can climb up onto, allowing them to gain access to the aviary above the rodent guard. As an added precaution, limit the amount of food available, remove spilt food every day and maintain bait stations with fresh bait.

Safety walkways are an ideal location for bait stations. This prevents the aviary occupants or animals such as pet dogs and cats from being accidentally baited. However, you should be aware that rats can and will carry bait around. Bait stations that prevent this from happening are available commercially.

A metal skirt around the aviary base can assist in rodent control.

FEEDING

Nutrition

Years ago, feeding your pet or aviary Cockatiel consisted of providing a bowl of dry seed and a bowl of water. Unfortunately, this boring and nutritionally inadequate diet is still fed by many people. Today, however, a better understanding of avian nutritional requirements and the introduction of a range of feeding alternatives such as pellets and dietary supplements are thankfully providing birds with a much more balanced and interesting diet.

To ensure the health and breeding success of your Cockatiels, it is important to understand their nutritional requirements.

A suitable dietary supplement for breeding Cockatiels. From left to right: quality Cockatiel seed mix, hulled oats, egg and biscuit, charcoal grit, multigrain bread, sprouted seeds and vegetables. Fresh water, cuttlebone and millet sprays are also provided.

Proteins—a combination of various amino acids—are vital for the overall growth of the bird. Muscles, eyes, skin, feathers and the nervous system all require adequate protein for their proper development. For breeding success the diet should include sources of protein that contain a balance of *essential* amino acids particularly lysine and methionine.

In captivity, it is difficult to provide the variety of foods required to attain the desired ratio of essential amino acids. Therefore, reliable sources of protein must be provided. Sprouted foods provide a good source of digestive enzymes and proteins, and are particularly recommended for breeding birds. Commercial pelleted diets specifically formulated for Cockatiels should contain adequate levels of essential amino acids.

Cockatiels also require a small percentage of essential **fatty acids**, especially omega-3 (linolenic) and omega-6 (linoleic) acid. While commercially grown seeds contain varying levels of fat content, many lack omega-3 and some have excess omega-6 fatty acids. Pelleted diets should contain a nutritionally balanced fatty acid composition. A small supplement of cold pressed linseed oil high in essential omega-3 can be provided for breeding birds.

Carbohydrates provide the primary sources of energy to maintain body warmth through their breakdown and supply of energy to all the functioning parts of the bird's body. Seed mixes including canary seed, French white, Japanese and Hungarian panicum millets are rich in energy. The importance of providing additional energy during breeding should be addressed by all Cockatiel breeders.

Vitamins are essential. Vitamin A is provided from plants in the form of provitamin A

A typical supplementary seed diet includes hulled oats, egg and biscuit mix and shell grit.

Diced fruits and vegetables and sprouted seed can be sprinkled with a nutritional supplement.

carotenoids—the precursors to vitamin A—necessary for good appetite, digestion, vision, resistance to infection and reproduction. Vitamin A is fat-soluble and therefore stored in fat deposits and the liver. Research into the level of vitamin A tolerance of Cockatiels has indicated that this species is vulnerable to vitamin A toxicity through their ability to store vitamin A efficiently. Even Cockatiels fed on a seemingly moderate vitamin A/kg^{-1} diet for 23 months were seen to be approaching toxic levels (Koutsos & Klasing 2005). A safe source of vitamin A is β-carotene, the precursor carotenoid found in foods such as carrot, red peppers, apricots, rockmelon, mango, sweet potato, leafy greens and particularly spirulina.

The B group vitamins help accelerate the recovery of ill and stressed birds. Vitamins A, D_3 and E also stimulate and restore the metabolism after a bout of illness. The required level of these vitamins in an all-seed diet is inadequate. Fruits and vegetables, particularly green vegetables, also provide vitamins and minerals, as well as valuable phytonutrients and digestive enzymes.

Vitamin D_3 is synthesised in the bird's body by exposure to ultraviolet light. It is important to note that birds are dependent on vitamin D_3 to maintain calcium levels in the body. To avoid calcium deficiencies, direct sunlight is by far the best source of vitamin D_3—essential for good fertility, strong eggs, proper bone development and other health benefits. However, the intensity of natural light in some areas of Europe and the UK are inadequate for maintaining vitamin D_3 and birds may therefore require artificial UVB lighting or supplementation.

Minerals, like vitamins, are an essential part of the diet. Calcium, phosphorous, iodine, iron, zinc and sodium are the most important minerals that birds need for proper function of the heart, muscles, blood and metabolism and are essential for bone development. Calcium is required in greater quantity by the egg-laying Cockatiel hen for quality eggshell formation, while a balanced phosphorus/calcium ratio is vital for bone formation.

Commercial vitamin, mineral and calcium supplements are readily available. Powdered forms are preferable to in-water supplements, as water intake varies with climate and individuals. Be aware that if you are providing a balanced pelleted diet additional supplements should not be necessary and can be harmful due to excess, particularly in vitamin A levels.

Breeder pelleted diets supply the correct proportions of vitamins, minerals and trace elements—probably one of the reasons that fertility is generally better in birds that are fed on pellets. Pellets also provide the additional energy required for successful breeding performance and good health.

It is debated whether **grit** is necessary in the diet, as very little appears to be consumed. However, insoluble grit aids digestion by grinding up seed in the gizzard and is particularly important for birds kept in pet cages or aviaries without access to sand and dirt floors. Soluble grit is digested by the acid in the proventriculus and therefore not used for grinding purposes. The soluble form is a good source of minerals and calcium—essential for general wellbeing and successful breeding.

Charcoal (top) and crushed eggshell grit.

A variety of grit mixes are available commercially. However, medicated grit is not recommended. The unnecessary medication of birds that are not unwell should be avoided as it can lead to the development of highly resistant strains of bacteria. Some grits contain charcoal which, while possibly beneficial, can absorb vitamins.

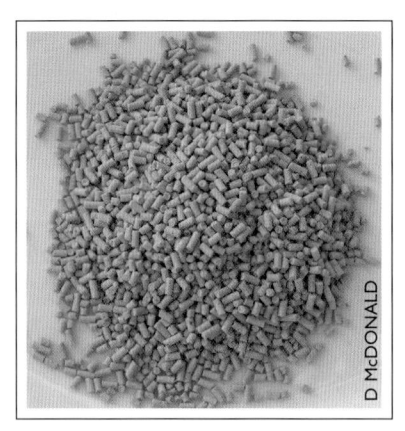

Pellets

Many brands of pellets are commercially available today. Most birds will do better on some form of pelleted diet rather than just seed, although some pellets have a higher nutritional value and some are more readily accepted than others. Pellets with a high vitamin A content are *not* recommended for Cockatiels, as they may cause toxicity and serious health problems. Recommended levels of vitamin A are in the range of 2000–4000 IU/kg^{-1} (dry matter basis). For this reason, you should choose only a pellet that is proven and specifically formulated for Cockatiels. Although pellets present a more balanced diet than seed, neither should constitute the entire diet of the birds. The recommended maintenance dietary ratio for a pellet-based diet is 80% pellets, 10–15% fruits and vegetables—particularly leafy greens—and 5–10% seeds.

Pellets provide a more nutritionally balanced diet than an all-seed diet.

The pelleted diet is extremely economical, due to the minimisation of waste. Because birds tend to crumble the pellets as they eat them and are not as fond of the fine powder that remains, feed only enough for the day so that they will pick up these crumbs and eat them, leaving only a few at the end of the day. Feeding a daily ration also helps control vermin as there is very little left to attract mice. Parents raising chicks should be provided with as much pelleted food as they require. Check their bowl in the afternoon to ensure that they still have enough pellets. Correctly formulated pellets supplemented with fresh fruit and vegetables have overcome many of the deficiency problems associated with an all-seed diet. Greenfoods, fruits and vegetables should be supplied in a different feeder.

Converting Your Birds to a Pelleted Diet

Some bird keepers believe that birds have to be forced to eat pellets, implying that there is an element of cruelty in feeding such a diet. It is true that you may often end up removing all other food sources in order to make the conversion, however, once they have accepted pellets as a food item—as opposed to something that you throw around as a toy—they will continue to eat pellets after other foods are reintroduced.

It appears to be a matter of recognition—birds need to recognise that pellets are something that they can eat. You may wish to wean young pet birds onto seed so that they will be able to crack and eat seed—in case their new owners do not feed pellets. However, young birds will usually begin eating pellets of their own accord as soon as they are offered. They sample new things all the time and rapidly learn that pellets are edible.

The following methods provide some creative ways of introducing pellets to birds primarily used to an all-seed diet:
- Mix similar quantities of seeds and pellets together. Add warm water until the mix is a sticky texture. Divide into rissole-shaped portions. Repeat the process daily—reducing the quantity of seed in the mix gradually over time.
- Sprinkle a small quantity of pellets over fruits and vegetables.
- Combine pellets into the seed mix, increasing the quantity of pellets over a period of time.
- Remove any toys from the cage and place a mirror on the floor. Sprinkle pellets over and around the mirror.

Case Study

Before converting my birds to pellets I fed a varied and successful diet that, for the majority of pairs, resulted in large healthy chicks. Although I produce the same quality of chicks on a pelleted diet, the growth rate is definitely faster and the chicks are more consistent in their development. The smallest chick that often becomes thin and weak on a conventional diet is a rarity, as the food being fed to the youngest is as nutritious as that being fed to the oldest. I have, however, encountered one problem. Due to the rapid growth rate of the chicks being fed on pellets, those that hatch later can often be squashed by the oldest chicks.

There are also great benefits for breeding birds—any egg-laying difficulties such as eggbinding or thin-shelled eggs are virtually eliminated. Improved fertility, hatchability, weaning and fledging are other benefits.

Diana Andersen

A quality Budgerigar seed mix is suitable for inclusion in a varied diet including fruits and vegetables.

Seeds

Seeds should only constitute a small portion of the diet. To satisfy your birds' nutritional requirements, fruits and vegetables, sprouted seeds and a variety of greenfoods should also be offered. In addition, a vitamin and mineral supplement may be recommended by an avian veterinarian for birds on a seed-based diet.

A high quality, small seed mix suitable for Budgerigars (in preference to a small parrot mix) is recommended. A quality Budgerigar mix usually contains plain canary seed, various millets and a proportion of hulled oats. There are seeds with a high fatty acid content in some small parrot mixes which are unsuitable for Cockatiels, ie sunflower and safflower seeds, which are high in omega-6 and have no omega-3 fatty acids. Most Cockatiels prefer to eat the smaller canary and millet seeds. Canary seed and hulled oats contain approximately 50% omega-6 and 2% omega-3 fatty acids. To increase the content of omega-3 in a seed diet add linseed (flaxseed) which has a 56% omega-3 fatty acid level.

You may prefer to mix your own seed so that it remains consistent throughout the year. A suitable basic mix should include 20% plain canary, 25% Japanese millet, 25% French white millet, 20% hulled oats and 10% linseed. Some sunflower (no more than 5%) can be added in cold climates during winter and for breeding pairs. This basic dry seed mix can be offered as a supplement to a pelleted diet.

This seed dispenser placed in a catching tray reduces the amount of discarded seed on the aviary floor.

This tray provides another method of reducing food spillage onto the aviary floor.

It is important to obtain good quality, clean seed from a reliable source. High quality seed should be of good size and contain little dust or husk. If you have any doubts about the quality and freshness of the seed, it can be tested by soaking and sprouting it. If seed is still viable, 95% of it should sprout within a few days. Seed that will not germinate or emits a foul odour is of little nutritional value and should be discarded. High quality seed mixes and commercial brands are available from reputable bird shops and supermarkets.

Seed must be stored in dry, clean airtight containers, preferably metal, to protect it from rodents, vermin, insects and sunlight.

Sprouted Seed

Sprouted seeds are richer in nutrients than dry seed. Breeding parents with young in the nest can be supplied with sprouted seed in virtually unlimited quantities. Other than breeder formulated pellets it is the best food for rearing chicks, being nutritious, bulky and easily digested. The mix should consist of one part grey sunflower, one part whole wheat, one part whole oats (horse feed oats) and one part white millet. Other suitable sprouting seeds are safflower, rape, milo, lupin (which is extremely high in protein), mung beans and pigeon peas.

Sprouts are more valuable as a dietary item if fed when the germinate part is approximately more than 1.5cm long. Acceptance of different types of sprouted seed may vary from bird to bird but on the whole it is a popular addition to the diet.

The seed selected for soaking or sprouting must be pre-tested for cleanliness by culture testing or sprouting on cotton wool. Seed that bubbles excessively, smells bad or grows fungus must be rejected.

Sprouted seeds such as this mix of mung beans, corn, pigeon mix and grey-striped sunflower seeds provide a good source of protein.

Sprouting needs to be done with care or the resulting fungal and bacterial growths within the mix can prove fatal to your birds. Poor soaking or sprouting can cause *E. coli* related health problems. An antifungal/bacterial agent should be used throughout the sprouting process, and the sprouts rinsed thoroughly prior to feeding. Grapefruit oil added to the water when soaking inhibits the growth of mould.

Sprouting Method

- Place the seed mix into a plastic mesh colander or sieve and rinse until the water runs clean.
- Empty the rinsed seed into a glass or stainless steel bowl and cover with a solution containing an antifungal agent at a ratio of 1:100.
- After 12–24 hours, tip the seed back into the colander and rinse thoroughly with water. Rinse the seed again by placing the colander into a larger bowl, allowing rapidly running water to lift and separate the seed until the water runs clean.

Three stages of sprouting seeds.

Thorough rinsing of sprouting seeds is essential prior to feeding.

- Immerse the colander into a solution containing the antifungal agent for 10 minutes, then lift and drain.
- Place the seed in a colander in a warm place.
- Repeat the rinsing procedure twice daily until the sprouts are approximately 1–1.5cm long and ready to be fed. This will vary, depending on the time of year, due to temperature changes. If you are using Aviclens™, the sprouts can be fed after the final dip into the solution. If you are using some other form of antibacterial or antifungal agent such as bleach, which is poisonous, rinse the solution off thoroughly with water before feeding. Abort the process if the seed has an offensive odour.

Depending on the nutritional balance of the staple diet, individual plates of sprouts can be sprinkled with a small amount of multivitamin powder prior to feeding. Sprouts should be served on clean plates separate to other foods each day and supplied to feeding parents as early in the morning as possible. All uneaten sprouts should be removed by the end of the day.

Fruits and Vegetables

Fruits and vegetables are a valuable source of vitamins and minerals, their colours, shapes and textures also providing Cockatiels with much enjoyment. The more leafy green vegetables that you can encourage your pet or breeding birds to eat, the better it is for their health, particularly if seed is the basis of their diet.

Suitable fruits and vegetables include spinach, endive, parsley, pumpkin, carrots, green beans, broccoli, silver beet, pomegranates and rockmelon.

Corn on the cob is particularly appreciated by feeding parents. This is not harmful if fed in small quantities. Corn has an unbalanced calcium to phosphorus ratio and has been associated with stunting and feather deformities in chicks.

When feeding silver beet, blanch it for 10 seconds in boiling water to break down the oxalic acid that it contains. The oxalic acid contained in silver beet is believed to prevent the absorption of calcium. Although there is some argument as to whether this is correct, parents readily feed softened silver beet to newly hatched chicks—it is obviously easily digested.

As fresh foods spoil rapidly, all uneaten leftovers should be removed at the end of the day.

Native Foods

Branches of eucalypts, with or without buds and flowers, acacia, casuarina, bottlebrush, melaleuca and lucerne (alfa alfa) are all enjoyed and

Native foods such as eucalypt, acacia, lucerne, wild oats and veldt grasses satisfy enrichment requirements.

recommended for Cockatiels. Other suitable bushes include those bearing berries such as hawthorn and elder.

The most obvious benefit of feeding non-toxic bushes is stress reduction. Branches will occupy the chewing urges of Cockatiels for hours, resulting in a lower incidence of behavioural disorders such as excessive preening, both of themselves and other aviary occupants. This is particularly important for pet Cockatiels that have little to occupy them in their confined cages.

Seeding grass heads provide a source of energy and enrichment.

Seeding Grasses

Regardless of whether you feed seed or pellets, seeding grass heads are particularly appreciated by all birds, especially parents feeding chicks. Wild oats, veldt grass, feed oats, sorghum and panicum are readily taken, even by birds unaccustomed to a varied diet. If these are unavailable in your area you can sow the seeds in garden pots. When seeding heads are about to open and drop their seeds, it is time to cut the sprays and feed them to your birds. They will attack the seeding heads with obvious pleasure. Lucerne (alfalfa) is also a highly nutritious, easy-to-grow perennial greenfood. For a complete list of suitable seeding grasses, refer to the table *Foods Eaten by Wild Cockatiels* on page 19.

Persuading Your Birds to Eat New Foods

A common complaint is that the birds will not eat what is offered. Educating your birds to enjoy a healthy variety of foods is a continuing process, but persistence pays off in the long term as birds that eat well tend to feed their chicks well. Young birds that have been fed a varied diet by their parents automatically eat a wide variety of foods.

The quickest way of educating your birds to eat new foods is to place birds that are unaccustomed to eating a varied diet with young birds that eat everything that is offered. If you only have one or two pet Cockatiels, introduce only one new type of food at a time, placing a fresh item in their cage each day. Eventually the birds become accustomed to it in their personal area and they will nibble at it. In doing so, they discover that this new object is edible. Small pieces or cubes of fruit or vegetable may be more inviting if offered on the end of a piece of wire or a cuttlefish clip.

Water

Dirty water is one of the primary causes of illness and death in aviary birds. Fresh, clean water should be provided daily in glazed or non-porous stainless steel dishes. Dishes should be removed at least twice weekly, more frequently in hotter climes, and cleaned thoroughly.

Water dishes should be located away from other food sources to prevent spoiling and contamination.

BREEDING

Selection of Breeding Stock

When allowed to choose their own mate, Cockatiels will form a strong pair bond and usually require no coaxing to breed. While this may seem ideal, it is rare that your choice of an ideal partner for a particular bird is also their choice. More often than not they form a bond with a sibling.

Line-breeding can develop the desired characteristics and establish good quality birds. However, inbreeding should be avoided. Brother to sister, mother to son and father to daughter combinations are all forms of inbreeding. In many cases this type of mating rapidly results in loss of fertility, loss of breeding urge, increased incidence of baldness regardless of the mutation variety, loss of size and generally produces less hardy birds. *Loss* is the overwhelming result of inbreeding.

Whiteface Suffused Grey (left) and Whiteface Dilute Cockatiel cocks.

Breeding pair of Normal Cockatiels, cock on right.

When breeding mutations, study the genetic inheritance and ideal pairings, taking into consideration the individual characteristics of the birds such as size, conformation, temperament, depth of colour and so on. It is important to be critical when looking at your birds. Identifying their strengths and their faults is the only way to improve your stock. For instance, a bird with a poorly shaped crest should not be paired with another bird with a poor crest. Often the desire to produce a rare colour overrides the decision to select a mate that will improve the conformation and strength of the offspring produced. This is understandable if numbers of the mutation colour are critically low but it would be advisable to outcross the resulting young as soon as possible—preferably to Normal birds with excellent conformation.

Gender Identification

While there are many factors that govern the successful breeding of any species, one factor remains common to all—the need for a cock and a hen. This may seem easy to determine

Page 42

Normal Cockatiel cock.  *Normal Cockatiel hen.*

when discussing the wildtype Cockatiel—a sexually dimorphic species. However, in captivity, there are a number of colour mutations. Mistakes in gender identification occur frequently because some mutations affect sexual dimorphism in both colour and feather pattern.

As a general rule, sexually mature, Normal Cockatiel cocks—and mutations that only involve a colour change, with no feather pattern change—have a bright, well-defined yellow mask, solidly coloured tail feathers and no juvenile underwing spots.

Mature Normal hens have tinges of yellow of varying intensity around the eyes and beak, and yellow undertail feathers, barred with grey. The orange cheekpatches are suffused with grey, making them appear duller than those of the cock. They also retain the underwing spots, although some birds occasionally fledge without these.

Young birds resemble mature hens and have a pink-toned cere and shorter tails and crests. At approximately four months of age, they begin a juvenile moult, which lasts about two months. At this time the cocks begin to develop their yellow faces. Some of their barred tail feathers are then replaced with solid colour. The hens produce new barred tail feathers.

Some families are slower than others to mature and changes to facial feathers may not be noticeable by four months of age. If this happens you can pull a barred tail feather to speed the process up a little. However, if the feather is pulled prior to four months of age, it may be replaced with another juvenile tail feather, giving you a false impression of a bird's gender.

Observed behaviour is another way of determining the sex of a bird. To our knowledge, no-one has ever witnessed the male shoulder-lifting courtship behaviour in a hen. (See *Courtship* on page 47.) If the colouring of the bird does not reveal its sex, this activity could be considered to be a fairly accurate way of identifying cocks. However, hens can sometimes sing as well as young cocks and I have also seen hens mating with other hens, which certainly complicates sex determination.

From left to right: Lutino Cinnamon Pied hen, Whiteface Cinnamon hen and Normal cock.

There are many other theories based on pelvic bones and cheekpatches, all of which are, as expected, at least 50% accurate. Some breeders opt for surgical or DNA sexing.

Colour Mutations

Lutino hens are expected to show yellow underwing spots on a creamy white background. However, this is not an accurate method of sexing Lutinos as it depends on the bird having enough yellow and white contrast for these spots to be visible. If the bird is also Pearl or Pied the normal wing spot pattern will be disrupted and this theory of sex determination will no longer apply. Lutino cocks develop a hormonal overlay at maturity that appears as a very pale brownish or mauve tinge to the wings that does not penetrate into the white wing bar. Hens never develop this colouration.

Cinnamon and Fallow mutation hens display considerably more yellow on the face than Normal hens. They do not, however, have anywhere near the amount of yellow that a cock develops nor does a hen of this mutation have the same clearly defined edge of the facial mask that the cock does. The cheekpatches of Lutino and Pied hens can be as large and bright as those of the cocks.

In Pied and Lutino mutations, behaviour—as described above—is often the best and possibly only indication of gender. A problem that arises when attempting to determine the gender of Pied birds is that you can only make the assumption that a bird is a hen

Lutino Cinnamon Pearl hen (left) and Cinnamon Pied Cockatiel cock.

Faded Cockatiel cock, known to Australian breeders as 'West Coast Silver'.

by the fact that you have not witnessed it behaving like a cock. There is always room for error if the bird is a slow maturing cock, or if you are unable to observe your birds all day for signs of male behaviour.

Breeding Season

Cockatiels will breed at any time of the year if given the opportunity. The fact that they will does not mean that they should, or that they will produce young of the same quality all year round. If your interest is in producing quality birds and maintaining valuable breeding stock, you can control the breeding cycle. The onset of spring generally triggers the breeding urge and this is also the safest climatic period for both parents and young.

After establishing breeding pairs, allow them to spend the winter together when the breeding urge is lowest. Winter is an ideal time to prepare the birds for the coming breeding season. In addition, pairs that are not immediately compatible will often form a bond by the time the season approaches. There is also enough time to introduce an alternative partner for any that are obviously incompatible. While the latter situation is rare, it does happen.

Although Cockatiels will breed in winter, there can be many disadvantages in doing so. The hens, particularly young hens, are more prone to eggbinding in colder weather. Any night fright will cause the total loss of eggs and young in a relatively short period and the shorter daylight hours will allow less time for the parents to feed the young. This results in chicks that are generally smaller, and that fledge later than usual. There is also a higher loss rate in younger chicks and a higher incidence of nutritional disorders such as calcium deficiency.

Provide a nestbox in late winter.

In light of this, compatible pairs should be given a nestbox late in winter. Hens should be watched closely for signs of eggbinding if they commence laying shortly thereafter. The benefit of this timing is that chicks commence hatching in spring when conditions are becoming milder and the hours of daylight increase as the feeding requirements of the young increase. In addition, a second clutch of young can be raised before the height of summer, avoiding possible losses in the nest caused by excessive heat in hotter climes.

If the location has a hot dry climate, remove the nestboxes after the second clutch. If a third clutch is desired, then the nestboxes can be returned to the aviary when the weather begins to cool. In cooler climates this may not be necessary. However, you have to decide if the quality of young produced in a third clutch by birds that have not had a break is worth the stress on the parents. It may not cause a problem for mature, experienced pairs in peak condition who are determined to breed.

Variations to these management practices apply, depending on the climatic conditions in your area. Some breeders prefer that their birds breed through winter and provide artificial light for up to 16 hours a day to extend daylight hours and eliminate many of the disadvantages of winter breeding. Extended lighting allows the birds extra eating and exercise time during winter.

With many species a suitable nesting site needs to be provided to stimulate breeding activity. Seasonal changes and the presence of a potential partner seem to be enough to stimulate most Cockatiels and if a nestbox is not provided, egg laying can commence with hens dropping eggs from the perch. This can become detrimental to their health as the normal trigger to cease laying eggs and incubate is absent, resulting in continued egg production. Unlike chickens who can lay many eggs, continuous egg laying in Cockatiels can lead to calcium deficiency, eggbinding, infection, prolapse, egg abnormalities and decreased embryo viability.

Colony Breeding

Being non-aggressive by nature, Cockatiels are one of the most suitable species for colony breeding, provided that you are not focussing on colour breeding. In a colony situation it is impossible to reliably identify the parents of any young produced. Hens will often share nests, resulting in chicks of mixed parentage being hatched in the same nest. Cocks may mate with more than one hen although they can only take up their incubation responsibilities with one hen. Genetically, the birds do not choose the best partner in order to produce the desired colours. There are other disadvantages. Nests containing too many eggs may cause low hatch rates. Hens working a nestbox that already has eggs in it may break the eggs. And occasionally, while parents are out feeding, another pair may take over a nest with young chicks, resulting in their loss.

In a colony aviary, nestboxes need to be placed as far apart as possible.

Colonies should be housed in aviaries with more width than holding or single-pair breeding aviaries to allow for nextboxes to be spread far enough apart to reduce the possibility of aggression between pairs. As a general rule you should increase the aviary dimensions by around 50% per additional pair.

Nestbox Preparation

Cover the bottom of the box with 5–7cm of firmly packed nesting material consisting of one-third coarse, untreated pine sawdust to two-thirds coco-peat, ie compressed coconut fibre that swells when water is added to form a sawdust-like material that is extremely absorbent. If the nesting material is too dusty it can be sifted through a 3mm screen to remove as much dust as possible before adding it to the nest. The main criterion is that the nesting material should be absorbent but not too fine and dusty as this can clog the nares and mouths of the chicks. Very dry materials, such as fine peat moss, are unsuitable because they absorb enough moisture to combat the hen's attempts to control the humidity for her eggs. Whatever you choose to use, ensure that your material does not come from wood treated with chemicals.

Nesting material should be absorbent and not too fine or dusty.

Once the nestboxes are provided, bonded Cockatiels are quick to inspect the nestbox. The cock is usually the first to enter and if he considers the box to be safe and suitable he will often be heard singing inside in an effort to entice the hen to enter. Part of the ritual of accepting the nestbox as a pair involves a lot of twittering discussion about remodelling and the entrance is often adjusted. However, Cockatiels are not usually destructive.

If pairs are not given enough time to bond before being given a nestbox, the hen may enter the box herself and commence laying and incubating before mating with the cock. If you observe that the cock is not taking his turn at sitting, this may be the cause. If there is a great deal of nestbox and mating activity by both birds you can expect eggs as early as one week after hanging the box. Considerable digging and scratching activity in the nesting material usually signals the onset of egg laying. (Also see *Nestboxes* on page 32.)

Courtship

Courtship is generally instigated by the cock. Healthy, active young cocks will commence singing and chortling to themselves at as early as eight weeks of age. Cocks will be seen lifting their shoulders while they flatten out their wings and cross the tips. While doing this they often appear to be doing a little 'soft shoe shuffle' from side to side, combined with repetitive singing. Some cocks raise and lower their bodies in a bowing motion to the hens while they lift their shoulders. Others follow the hen of their choice endlessly around the aviary, making little hops or skips as they go. The hen

Courtship behaviour by a Platinum Pearl pair.

usually appears oblivious to the cock's attentions unless she wishes to mate, then she crouches down with her tail elevated, and makes a twittering sound. If nesting sites have been made available, this behaviour is generally accompanied by frequent visits to the nestbox or log of their choice where both birds enter and prepare the nest.

Egg Laying and Nest Inspection

Cockatiel eggs are white and normally about 25mm long. Four to seven eggs are usually laid on alternate days. However some young hens will occasionally continue laying until the chicks hatch.

Try to accustom your birds to regular nest inspections and do so by talking quietly to the pair in the nestbox, while gently rubbing on the inspection hatch before opening it. The birds often depart but some stay in the box moving to one side or rocking and hissing above their eggs. This is ideal if they calmly allow you to handle the eggs, but parents who depart in fright when your hand enters the nest can scatter and dent their eggs. Try to slide your hand in so that the back of your hand is towards the adult bird and the eggs are shielded from damage if the bird becomes alarmed and leaves suddenly. After a few regular nest inspections, most fully mature birds will accept the inspection calmly.

One benefit of nest inspection during egg laying is that it allows you to mark the eggs as they are laid and to candle the eggs for fertility. You will be able to predict exactly when the eggs should start to hatch. If a hen lays excessively, she will not be able to incubate all the eggs properly. By labelling the eggs with a non-toxic marker in the order that they were laid, you can remove the last eggs as they appear, provided that you have candled the

Eggs and newly hatched chicks.

first 5–6 eggs and found them to be fertile. It also allows you to recognise potential egg-laying problems. If eggs do not appear when they are due or the eggs are abnormal in some way, observe the hen closely for signs of poor health.

Incubation

Incubation lasts 18–21 days depending on when the pair begin to sit tightly and how warm the days are. Typically, both parents begin to incubate after the hen lays her third egg so that up to three babies can hatch on the one day. However, if temperatures are very warm, incubation can start almost immediately. The hen usually incubates at night and the cock during the day. Particularly enthusiastic hens will sometimes do both shifts and very devoted pairs will often incubate together. The cock does not feed the hen so both the hen and the cock will come off the nest from time to time to feed.

Most young birds can be educated to be good parents with a little foresight and patience. It is better to try and educate your birds to be good parents than to train them to lay eggs for artificial incubation production or to give up and get rid of the pair. Always expect first-time pairs to be less successful than older, experienced birds. Sometimes first nests are laid, incubated, fed and fledged perfectly. Birds that are closer to two years of age instead of 'about' one year of age when they attempt breeding are usually more successful with their first nest.

Incubation Problems

Unfortunately, when it comes to breeding, things do not always run smoothly and all too often our efforts to rectify the situation only make things worse. Below are some common problems encountered during egg laying and incubation and some possible remedies.

Infertile or Clear Eggs

If a young pair who breed for the first time lay and incubate perfectly but do not produce fertile eggs, replace a few of their clear eggs with fertile eggs from another pair. If they hatch and raise the chicks together successfully, their next clutch of eggs is usually fertile. Their hormone cycles seem to settle down and be synchronised for the next attempt. If you remove the infertile eggs in order to stimulate them into producing another clutch of their own, you may alter their brooding instinct and cause other incubation problems.

Infertile eggs can also result from eager hens laying eggs before bonding and mating have occurred. If you suspect that this is the case, remove the eggs and the nestbox for several weeks. This will allow the pair adequate time to become acquainted.

Collect any clear eggs on a regular basis to have a supply to use in emergencies. Old eggs generally dry out and lose weight and are not suitable to use. Some specialist pet or bird dealers have very realistic polyurethane eggs that are perfect for substitution. These artificial eggs are similar in weight and size to Cockatiel eggs and retain their brooding temperature just like normal eggs.

Cockatiel eggs in the nest.

Above: Various stages of developing chicks.
Below: The eldest in this clutch of Whiteface Cinnamon Pied and Whiteface Pearl Pied chicks is 26 days of age.

Hens That Fail to Incubate

This is a common problem with first-time hens. They are best paired with an experienced cock that will incubate his shift regardless of the hen's activities. These hens often lay continuously because the lack of brooding instinct that usually terminates egg laying has resulted in the continuation of egg production hormones. If these hens are entering the nestbox to lay and to sit intermittently with the cock, then a pipping egg from another pair can be placed in the nest while the cock is sitting. Again, the realisation that young are hatching often rectifies the situation and the hen may have no further problems. If the cock has not incubated during the night the eggs will have died. If they are important eggs, place clear eggs under the pair and artificially incubate the original eggs, placing them back under the pair as each one pips in the incubator.

Incubating hen.

Inexperienced hens sometimes leave the nest prior to the completion of incubation. Try replacing some of their eggs with eggs that are more advanced than their own. These will commence pipping while the hen is still incubating. The realisation that young eventually hatch from the eggs will usually trigger the normally reliable Cockatiel maternal instincts.

Cocks That Fail to Incubate

This problem is a little more difficult to correct because the cock has no reason to enter the nestbox except to check on the hen. However, if you leave the hen sitting on clear eggs and artificially incubate the fertile ones, you can try hatching an egg under the hen. This is difficult, primarily because many chicks hatch in the very early morning when the cock is due to take over. If the hen leaves the nest at daybreak and the cock does not take over, the chick could rapidly die from cold. If you want to try this exercise, you need to get up at daybreak and be ready to rescue the chick if the cock does not assume his fatherly duties.

Usually the hen incubates at night and the cock during the day.

I have had two cocks that do not incubate until they hear the hens feeding young. Then they become perfect fathers, brooding their young through the day and incubating and hatching the remaining eggs. Some hens will raise a clutch on their own although this is very difficult for the hen who should not be left with a large number of young to raise alone.

PARENT REARING

After the initial pip in most trouble-free hatches, a chick emerges within 48 hours. If the chick takes longer to emerge, you should watch closely for signs of difficulty. Hatching difficulties are more likely to occur in hot dry weather, because the internal membrane of the egg dries out. Hens regularly walk into their water dish in order to transport moisture into the nest. In very dry conditions she may not be able to maintain the humidity at the required level. This lack of moisture will also cause the internal membrane to dry out and the chick will be unable to rotate freely to complete chipping out of the egg. If you suspect that pipping eggs are drying out you can moisten them morning and night with a clean cotton ball and warm, boiled water.

Chicks that have taken a long time to emerge or have been assisted in hatching are often dehydrated and weak and have little chance of survival if they remain with the parents. It is best to remove them for handrearing for 24–48 hours until they are strong enough to be fed properly by their parents. When they can hold their head up to beg for food they can be safely returned to the nest. After hatching, all chicks should be observed in the evening and for the next few days to ensure that they are strong enough to survive. No matter how good the parents are, they can do little to help a chick that is unable to hold its head up to feed.

At hatching, remove the shell and any membrane and place the chick in a brooder. Note the egg tooth.

If a chick has some of the yolk sac still attached to the body after hatching, remove the chick immediately for handrearing. Do not pull at or remove any of the yolk sac. Instead, treat the area gently with a cotton bud and Betadine® solution several times a day. In many cases, if the chick is kept on a clean soft surface such as facial tissues and infection is avoided, the yolk sac will eventually dry up and fall off.

Many people believe that weak chicks should not be assisted to survive. However, if chicks are weak only because of environmental factors in the nest then it is a waste to let these chicks die. Assisted chicks that were experiencing difficulty can grow into exceptional adults.

Chick Development

Chicks grow rapidly for the first three weeks, often reaching two-thirds of their adult body weight by the time they are two weeks of age. Supplying insufficient food to feeding parents is one of the most common mistakes made by breeders. It is remarkable how much food the parents require to meet the demands of their voracious young. Ensure that there is more than the parents require by checking their food supply morning and afternoon.

Pin-feathers begin to show at as early as seven days of age and chicks should be fully feathered by four weeks of age. Feather development appears to be governed by the growth rate of the chick.

Above: Two Whiteface and Normal (showing yellow down) chicks at 3–5 days of age.
Below: Whiteface Cockatiel chick at 11 days of age.

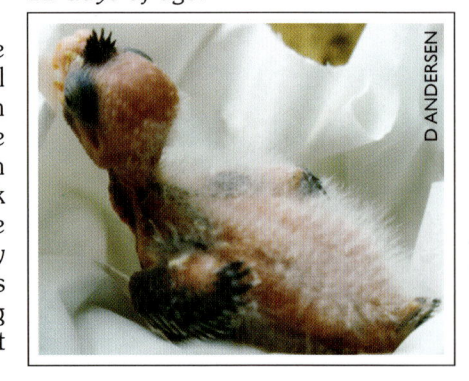

Monitoring Growth

Often the youngest chick will not require banding nor show signs of pin-feathers until two weeks of age, taking a full week longer than the oldest chick to reach the same stage. These chicks should be closely monitored, as growth abnormalities and stunting can result in a chick that falls too far behind. Handraising these chicks may help them survive but it can rarely reverse the damage caused in the early stages if they are removed too late. If a slow-growing chick is recognised before suffering permanent

Normal chick at 11 days of age. Note the yellow down and the developing pin-feathers.

Whiteface Cinnamon Pearl Cockatiel chick at approximately three weeks of age. Note the white down and absence of the cheekpatch in this mutation.

Chick developing pin-feathers.

stunting, the feeding of a good handrearing diet will help it catch up to its nest mates.

Handrearing can sometimes be avoided by transferring young from one nest to another. This should only be done where accurate records are kept and identification of the young is in no doubt. For instance, if a pair hatches a fifth and sixth chick, these often get into difficulty within a short period of time. They will either die or require handrearing, as they are too small to compete with the older chicks for food.

If you have several pairs breeding at the same time you can often place the two youngest from a nest of six with the youngest from another nest, removing two of their older chicks. The more advanced chicks can be placed with the older chicks from the first nest. If all the young birds in the nest are about the same size they can compete more effectively for food.

One disadvantage of this practice is that it can expose chicks from both nests to carrier diseases which are not evident in the parents. New chicks added to the nest may have no immunity and therefore quickly contract the disease. For the same reason, avoid handling chicks between nests without either washing hands thoroughly or wearing disposable latex gloves.

Closed-banded chicks can be easily identified but unbanded chicks must be marked clearly so that there is no question about their origin. Their rumps can be marked with a non-toxic water-based marker, although this needs to be checked morning and night and renewed as necessary until the chicks are old enough to be close banded. Any movement of chicks from one nest to another needs to be done gradually, ie one at a time, so that you can be sure that the chick has been accepted and that the parents do not abandon their young.

Chicks that appear to be falling behind can also be assisted by filling their crops first thing in the morning with a good handrearing mix. Small chicks are weakest when their crops have been empty for some time. They are also the last to be fed, sometimes waiting several hours while the parents feed the oldest chicks first. This weakens them even more until they may be too weak to ask for food. Giving them a feed will give them the boost that they need in order to compete for food and will at least sustain them until their parents get around to feeding them.

Feeding chicks in the nest can be a slow and painstaking operation unless you are competent with a crop tube. They are not used to a spoon or syringe, so the food needs to be quite runny and warm to get them to accept it. As you will rarely get a normal feeding response, you need to feed small amounts to reduce the risk of the chick inhaling the mix and aspirating.

By the end of each day chicks should have full crops but the contents should feel neither excessively dry nor excessively wet. Either of these conditions could signal a potential problem with the chick. As the chicks approach four weeks of age, the amount of food found each evening in their crop begins to reduce as the chicks slim down in anticipation of fledging.

Physical Check

My birds are all accustomed to regular nest inspections and, when young are in the nest, chicks are checked at least once a day. Droppings build up rapidly in the nest and occasionally a chick will get its vent or beak blocked by dried droppings. Such blockages can lead to a quick death. Therefore, a quick 'top and tail' check is recommended to ensure that both are clear. A chick that has a pasted vent needs to be observed closely for signs of possible diarrhoea that may have either caused or resulted in the problem. A blocked vent can rapidly cause constipation and kidney failure. A chick that has its beak clogged with droppings may have been without food for many hours and may need handrearing formula to help it regain its strength. (See *Handrearing* on page 58.)

Chicks should also be monitored for normal leg development. If your breeding birds are supplied with a nutritionally balanced diet providing a plentiful supply of calcium and nestboxes have adequate nesting material, this should rarely be a problem. However, any abnormalities, such as splayed legs or an injury, need immediate attention if they are to heal adequately.

It is usually best to remove a chick with leg problems for handrearing. If the problem is in the early stages, place the chick in a very small container that does not allow the legs to slip out sideways. If this is not successful, cut a thin strip of electrical insulation tape, wrap it around one leg and then across to the other leg. Loop it around the second leg, pulling the legs closer together but wide enough apart for the chick to sit comfortably with its legs parallel below it. Stick the end of the tape to itself between the legs. Insulation tape does not stick well to the skin but it does stick well to itself, the tape acting as a set of hobbles without restricting blood flow or growth in the ankles. Due to the rapid rate of growth in young chicks, the problem can be rectified within a few days if attended to early enough. If the condition is advanced or involves an injury, consult an avian veterinarian.

Feather Development

Feather development is another good way of monitoring the health of a chick. Cockatiels develop quickly in the nest and are often weaned while other parrot species of comparable size are still in the nest. As a result, the contour feathers develop before the down feathers. Other species such as lorikeets develop down feathers first, giving them the appearance of wearing flannel pyjamas. The development of these contour feathers should be well under way by three weeks of age and complete by four weeks of age, in preparation for fledging. The tail may not be full length at fledging but all other feathers should have finished growing.

If a chick is stunted or unwell, feather development will be delayed. Stress lines may also appear on the opening feather, indicating insufficient nutrient absorption during the growth period. Under no circumstances should feathers fall out as they are

Young Cockatiels displaying healthy feather development.

developing unless it is the result of an obvious trauma causing the feather to die. In this case a new healthy feather will begin to grow in its place.

Fledging

Young Cockatiels will usually be independent by 7–8 weeks of age but should not be removed from their parents until all signs of begging have ceased. Once the young have fledged they need to be watched to ensure that they are still being fed and that they have sustained no injuries in their initial 'crash bang' efforts of flying. Chicks will appear to be husking seed shortly after fledging, but in many cases they are only imitating their parents and are not actually cracking or swallowing seed. (See *Weaning* on page 65.)

Feather Plucking

Feather plucking in the nest is probably one of the most frustrating problems encountered by Cockatiel breeders. Unfortunately there are no simple solutions to breaking this habit. In some cases feather plucking is a learnt habit. Most birds that have been plucked as chicks will, in turn, pluck their own young. If you remove the chicks for handrearing as soon as plucking becomes evident you can certainly reduce the likelihood of future plucking. However, parents that have never been plucked will often, without warning, begin to pluck their chicks. Some will raise several clutches before the habit begins.

There are several possible causes of the habit. In the past, one of my hens plucked the heads of her chicks when she began to lay and incubate another clutch of eggs, before the previous clutch had fledged. While brooding this second clutch the weather became quite hot so she was not inclined to lay another round of eggs. Therefore, her next clutch fledged and weaned without a feather out of place.

Other parents attack their chicks quite viciously without warning, removing many feathers and often inflicting a number of small wounds in a single day. I am sure that this is done in an effort to drive the young out of the nest. One solution involves placing another nestbox in the aviary for the hen to use while the chicks are fledging and hope that the cock will continue to feed them. This is not always successful, especially for Cockatiels, because the cock would normally be in the nestbox incubating during the day and providing another nestbox for the hen assumes that she will vacate her chosen nestbox. Most hens seem to want to use the same box.

Dominant Edged Cockatiel fledgling.

Above: These Cockatiel chicks have been removed from their parents who tend to feather pick their young.
Below: Feather plucking is in some cases a learnt habit.

Hyperactive birds seem more likely to pluck, particularly hens that are required to brood their chicks for long hours during the night. Chicks with light-coloured feathers seem to be more commonly plucked. This may simply be because their pin-feathers are more visible in the dark.

There may be a multitude of other causes that we are unaware of. At the moment the only solutions that are effective to any degree are the provision of a good diet including minerals and trace elements, reduction of stress-related environmental factors and breaking the feather-plucking cycle through handrearing.

ARTIFICIAL INCUBATION

Although it is important to encourage breeding birds to be good parents, it is advised to have an incubator available throughout the breeding season. For a variety of reasons, many eggs pass through the incubators and quite a few birds only survive because of them! However, many eggs are returned to the nest at the point of hatching, resulting in the need to handraise fewer chicks from the egg.

When breeding mutations you often have only younger, inexperienced birds to use, and the incubator is helpful in sorting out parent incubation problems. Eggs will not be lost through irregular incubation and can often be hatched under the parents, solving the problem on a permanent basis. Some hens lay thin-shelled eggs that are frequently damaged. Their eggs can be brought to hatching in the incubator and then returned to the hens at hatching, if clear or artificial eggs have been left for them to continue their natural cycle of incubation.

Chicks with hatching difficulties can be watched more closely in the incubator and assisted in hatching if necessary. These chicks can remain in the incubator while being fed until they are strong enough to be placed under the parents. If necessity dictates, I will handraise from the egg but only when no alternative can be found. It is a long and tedious task that can be both rewarding and terribly frustrating.

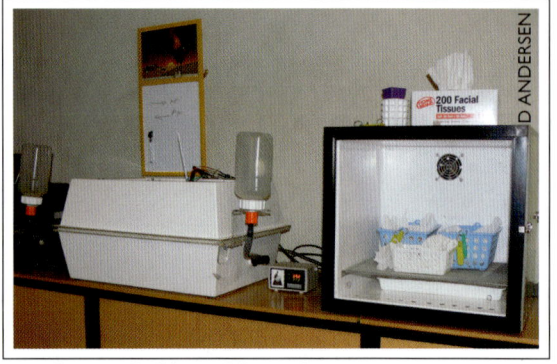

Two fan-forced, automatic turning incubators are recommended. One incubator can be used for eggs developing from day one to pipping. The ideal temperature is 37.5°C with a humidity level of 50%. The automatic turning device turns the eggs 90° every two hours. Eggs turn a total of 180° in one direction and then 180° in the other direction, completing a 360° rotation every eight hours. When turning eggs by hand, care should be taken to ensure that eggs are turned equally in opposing directions otherwise the internal egg structure will be damaged.

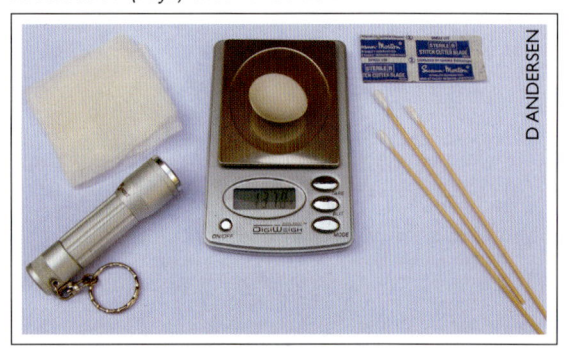

If you are incubating eggs regularly or breeding colour mutations it is important to keep the

Incubator (left) and brooder.

A candling torch and egg weighing scales are necessary for monitoring egg development.
In case it is necessary to assist a hatch it is beneficial to have a sterile scalpel blade, sterile swabs and some clean gauze to rest the egg on.

different batches of eggs separate. To ensure that the genetic background of the eggs is maintained, they should be marked carefully with a pencil before being placed in the incubator. When young hatch, they should also be clearly identified. This can be done by separating the chicks into different baskets in the brooder that are labelled with the parental details. I have used plastic luggage tags that can be washed along with the baskets. However, any form of labelling that allows you to identify the container and its occupants will suffice.

Chicks are more content when kept in groups. Therefore, if an individual chick needs to be identified so that it can be added to a group, the skin can be marked lightly with a non-toxic marker. As this tends to wear off within a couple of days it needs to be redone before it fades. Once the chicks are banded and their details recorded, identification is no longer an issue.

Hatching

Eggs should not be turned for the last three days of incubation or after the internal pip has commenced. The pip can be determined by candling, or by waiting until the first dent appears on the outside of the eggshell. At this point, transfer the egg to a hatching incubator. This has a much higher humidity level and a slightly lower temperature. The automatic turning mechanism needs to be disengaged.

The humidity level should be at least 65%, slightly higher if possible—with a temperature of 36.9°C. In the hatching process the chick initially breaks through the membrane into the internal air cell when it begins to breathe air. This is referred to as internal pip. The chick then dents the external shell which allows fresh air into the egg and the membrane begins to dry. The chick appears to work on one area for a minimum of 24–48 hours while the remaining yolk sac is drawn into the body. A chick that is strong and progressing normally can be heard tapping away inside the egg and usually begins to cheep on the second day.

If the chick is hatching normally, it will begin to chip the shell in a circular pattern as it rotates in the eggshell after the yolk absorption process is complete. The chick breaks through the shell until the end pops off, then struggles out, kicking off the remaining shell. The rotation process should take no more than about 10 minutes. It is disastrous to open an egg before the yolk sac has been absorbed. However, it is rare for the rotation process to begin before absorption is complete. If you feel that the chick is making little progress after rotation has commenced, it is possible that it is stuck to the membrane—it is best to assist the hatching chick at this point.

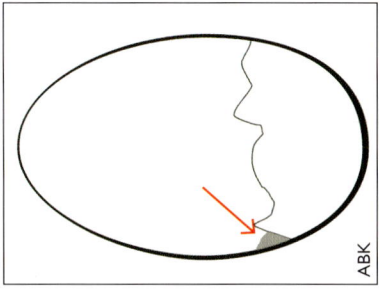

Above: A blurry shadow at the bottom of the egg is often the first sign of internal pip.
Below: This egg is in the rotation stage prior to hatching.

Incubators can be used to increase the yield of rare or valuable birds. However, the practice of continuously removing eggs can result in future breeding problems for a pair that may otherwise have been good reliable parents. I have bred many incubator-hatched, handraised birds successfully, but never allowing Cockatiel parents to raise their own young may result in generations of birds that lack the skills to do so. On the other hand, I rarely have problems with any birds that have been hatched and raised by very good parents. They seem to inherit their parents' skills.

Brooding Conditions

The younger the chicks, the more crucial temperature and humidity are. Tiny chicks are unable to tolerate temperature fluctuations and need to be kept in a stable environment. Although an incubator will suffice while the chick dries out, in a fan-forced unit the chick may become dehydrated if it remains in an environment where the air is blowing directly onto it. With advances in computer technology extremely accurate brooders with a digital temperature control are now available. This makes life much easier in terms of maintaining temperature control. Fan-forced brooders are suitable, provided that the air does not blow directly onto the chicks. A well-designed brooder should have accurate temperature control, adequate air circulation and be easy to keep clean. If you are able to clean and disinfect your brooding equipment thoroughly it will greatly reduce the incidence of bacterial and viral disease transmission.

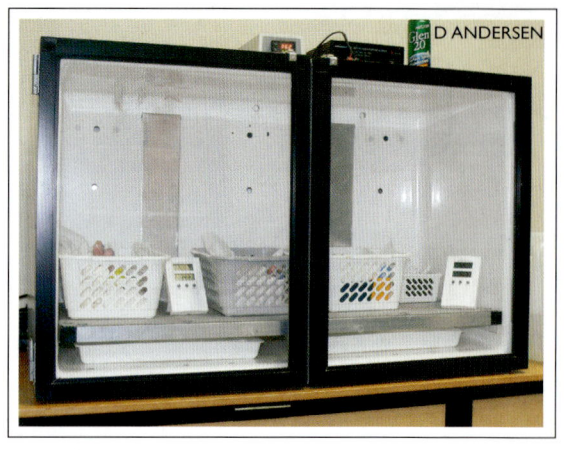

Brooders set up with chicks and humidity readers.

Recommended brooding temperatures are listed below. (Note: parent-raised chicks that are removed from the nest at two weeks of age or older often seem to find these temperatures too high.)

Week 1   35–37°C
Week 2   32.5°C
Week 3   27°C

The temperature in the brooder should be gradually decreased over a 48-hour period. The behaviour of the chicks and their droppings, as well as the temperature, must be monitored closely. Chicks should not shiver, nor should they sit apart from each other and pant. Between feeds contented chicks will normally sleep for hours in a little heap with their necks crossed over each other. The youngest chick is always cuddled up underneath and in the centre of the warm bodies. It is much more difficult to keep an individual chick content as it has no-one else to snuggle up to for warmth and comfort.

Plastic tags are used to identify the chicks in the brooder.

If temperatures are too high at the stage when the handrearing formula is gradually thickened, dehydration can occur. The first signs of this can be seen in the chicks' droppings. Normal droppings should be well formed but not dry. If droppings are not accompanied by a reasonable amount of liquid, monitor your brooding conditions carefully.

Homemade Brooders

If the chicks are more than one week of age you can manage with a hospital box fitted with two 25-watt bulbs and a dimmer switch. The benefit of having two bulbs is that the chicks will probably survive the drop in temperature if one bulb fails. On the other hand, if there is only one bulb in the box and it fails, the chicks can rapidly die before the problem is noticed. Dimmer switches do not prevent the temperature in the box from gradually rising and declining as the room temperature fluctuates throughout the day and night. Without a thermostat the heat level will require constant monitoring and adjustment.

Thermometers are best placed in the container that holds the chicks as sometimes conditions inside the container can vary several degrees from the environment outside the container. Handy, inexpensive digital thermometers that have a probe in the end of a long cord are readily available. The probe can be easily positioned in the container while the thermometer itself remains outside the brooder for easy reading. Ensure that the chicks are not actually leaning against the probe itself or you will have abnormally high readings.

Older chicks beginning to pin-feather are much easier to cater for. These chicks are best housed in plastic storage boxes. Insulated coolers are good for smaller chicks that require warmer temperatures. The containers should be large enough to position a heat source on one side. The chicks can then move closer to or further away from the heat source as required. A towel can be placed over the top, leaving the top end opposite the heat source open. To ensure that the environment is within an acceptable temperature range a thermometer should be placed in the box. The box should have a firm base, eg shredded paper covered with paper towel. Such an environment is very easy to clean and disinfect.

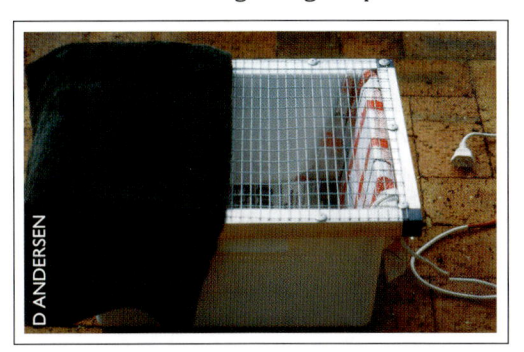

A plastic storage box fitted with a heat pad at one end is a suitable homemade brooder design.

Plastic heat pads—available from pet suppliers—are recommended as the heat source for chicks in this type of brooding situation. The heat pad can usually be positioned vertically at the end of the brooder box. Chicks can lean up against the pad or move away from it if desired. Although this type of heat source is very gentle, cover the heat pad with a small, clean handtowel or pillowcase to prevent chicks burning the exposed skin on their crops. This also keeps the pad clean. Another benefit of these pads is that, being plastic, they can be wiped over with disinfectant. Light bulbs can also be used as a heat source but they cannot be easily cleaned and are more inclined to burn chicks if too high a wattage is used and the bulb is not isolated from direct contact with the chicks.

Humidity Levels

Any brooder that you use requires humidity to prevent dehydration in the young. This can be provided by placing a dish of water close to the heat source. Although a dish 8–10cm in diameter is usually adequate, it depends on the design of the brooder and how high the required temperature is. The higher the temperature is in a brooder, the greater the surface area of the water dish needs to be in order to maintain the required humidity level. With older chicks kept in plastic tubs there is generally enough moisture being produced by their droppings at this stage not to require additional humidity. However, it is still necessary to monitor your chicks for signs of dry skin and dehydration.

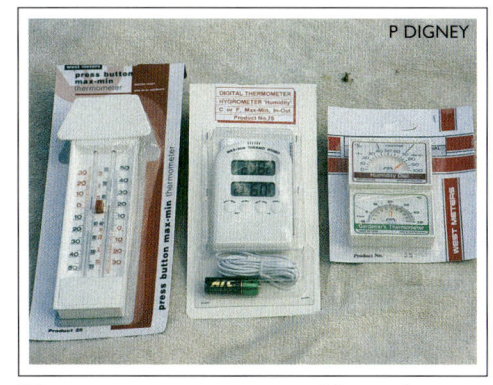

There are now a variety of humidity readers available. The digital models have the advantage of recording high and low variations in your absence.

Maintaining Hygiene

Regardless of the age of the chicks, poor hygiene will ultimately result in problems. Incubator-hatched young do not have the benefits of their parents' antibodies and digestive enzymes in their crop. Until the chicks develop their own immunities they are highly susceptible to bacterial infections. For this reason incubators and brooders require regular cleaning and disinfection and the nursery itself needs to be kept as clean as possible.

Keep chicks that are beginning to feather in a different room from incubators and brooders containing very small chicks. Dust from discarded feather casings is produced in large amounts and spread enthusiastically by chicks practising to fly. Not only is the dust a health hazard to very young chicks but it is also potentially damaging to the sensitive mechanisms of incubators and brooders.

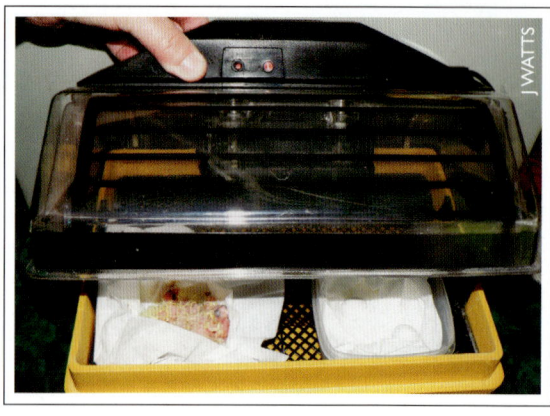

Containers housing chicks and eggs should be lined with clean, soft facial tissues.

Containers housing small chicks should be lined with a bed of creased, soft facial tissues. For larger chicks, a bed of shredded copy paper covered with paper towel is suitable. It is important to provide the chicks with bedding that their legs can grip on to. Surfaces that are too flat and smooth can result in chicks with splayed legs, as can large containers that allow the chicks to move around too much. Therefore, tiny chicks should be placed in small containers that restrict their movement. This can also be achieved by packing scrunched-up facial tissues around the chicks.

After each feed the container linings should be changed. The containers should also be replaced each day with clean disinfected ones. Older chicks removed from the nest may have a higher resistance to pathogens and therefore may not require such strict hygiene. However, it is a good practice, which will reduce the level of bacteria that may colonise your nursery.

HANDREARING

All bird keepers will have to handrear chicks at some stage in their hobby. There are many reasons for this, eg increased production; assisting weak, slow-growing or feather-plucked chicks; or merely to imprint young birds as pets. Although time consuming, it need not be a difficult process provided that several essential requirements are catered for adequately.

Assessing the Condition of Chicks

At hatch, Cockatiels are covered with soft yellow down. The only variation to this is the Whiteface mutation whose chicks hatch with white down. This is consistent with the effects that the mutation has on adult feather colouration.

Shortly after hatch—before the chick's down has dried and become fluffy—is a good time to assess the condition of the hatchling and determine

At hatch Normal Cockatiels are covered in yellow down and Whiteface Cockatiels are covered in white down.

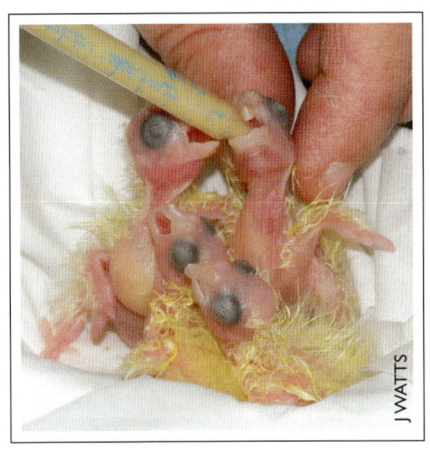

These two-day-old chicks display a healthy, plump physique with pink skin tones.

if there are any problems that may need to be addressed. Chicks should be plump and have pink skin tones. Chicks that are reddish and have a skinny appearance particularly in the extremities, eg wingtips and toes, are likely to have become dehydrated due to incorrect incubation techniques or a prolonged hatch. Unless these chicks are properly re-hydrated they are likely to progress slowly and show signs of stunting if they do not die soon after hatching.

Chicks can exhibit signs of dehydration at any stage—a condition that needs immediate attention. More often than not, a nestling that is not progressing will be dehydrated, sometimes severely. However, chicks that are extremely pale may be carrying a disease or a bacterial infection or may have haemorrhaged during hatch.

Some breeders believe that newly hatched chicks should not be fed for up to 24 hours after hatching. This advice is suitable only if the youngsters hatch normally with no problems. Chicks that have had a difficult hatch often need immediate help. They can be safely fed an electrolyte/glucose solution for the first 24 hours, using Gastrolyte™ (available from chemists) or solutions made specifically for birds, eg Spark™ from Vetafarm. If you are able to obtain them, intravenous packs of Hartmann's™ solution and glucose can also be used. Given orally, a 50/50 mix is suitable. This solution gives the chicks the energy they require without interfering with the remaining internal yolk absorption. It is also of great benefit to chicks that have become dehydrated in the egg. For these chicks I continue adding electrolytes to their formula until the dehydration has been corrected.

One of the benefits of using intravenous electrolytes is that the solutions are sterile. All you need to do is draw out the required solution with a sterile needle. If using a powdered electrolyte, the solution needs to be discarded after about three hours. If you use extremely accurate electronic scales you can mix it fresh for each feed without any waste.

Handrearing Formulas

I have the utmost respect for anyone who successfully handraised chicks prior to the introduction of the range of commercial avian handrearing formulas that are now available. It was certainly more difficult to handrear young Cockatiels than it is now. There are several excellent formulas on the market that take all the guesswork out of feeding chicks. They are highly digestible, totally balanced diets requiring nothing more than the addition of water and heat. When fed according to the manufacturer's directions, you should be able to raise youngsters with weaning weights equivalent to, and in some cases, better than those of parent-raised chicks.

The handrearing formula should always be mixed fresh with cooled, boiled water for each feed. Use separate batches of food and fresh feeding implements for each group of chicks. In this way a bacterial, viral or fungal problem can be prevented from spreading through your nursery before it is detected.

The formula will lose nutrients and turn gluggy if mixed in extremely hot water.

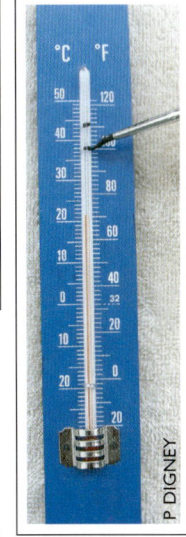

Mark the temperature with oil-based paint.

The most important rule with commercial foods is to feed them as per the manufacturer's directions. Many people mix one brand with another and add different ingredients that may ultimately result in digestion problems and will certainly alter the guaranteed nutritional value. While some species of parrots may need additives in order to attain optimum growth rates, we have always handreared Cockatiels successfully without altering the formula.

A set of accurate scales for weighing the ingredients makes it easier to mix the correct consistency of formula relative to the age of the chicks. Inexpensive, digital kitchen scales, which read in 1–2 gram increments, are available from electrical appliance stores. Scales that are accurate to 0.1 gram are far better if you are raising tiny chicks that require very dilute formula but the cost of these can be prohibitive.

Handrearing formulas differ in digestibility and some produce better weight gains than others. If the formula is not mixed to the correct consistency or not enough care has been taken with hygiene, this may lead to digestion problems. At the first sign of a slowing digestion a more dilute mix should be fed. If this alone does not rectify the problem, Nilstat™ antifungal drops (for human babies) should be administered into the crop, preferably when the crop is empty or close to empty, until the problem has cleared. If the problem is not quickly resolved an avian veterinarian should be consulted. The longer a digestion problem is allowed to continue, the weaker the chick becomes and the less likely it is to make a full recovery.

Place the thermometer into the centre of the formula for accurate temperature measurement. Placing the formula in a bowl of warm water helps maintain the temperature.

Chicks prepared for handrearing.

The temperature of the formula is also extremely important. To a jug of boiled water that has been allowed to cool add boiling water to a temperature of around 48°C. This is then added to the formula. Never add water that is boiling or close to boiling to the formula. With some mixes this has the effect of cooking the mix and altering its qualities and digestibility. Test the food with a thermometer and feed as close to 40°C as possible. Formula fed over 42°C can cause crop burn, while formula fed too cold can chill the chick or cause it to refuse to feed. If you cannot test the formula with a thermometer you should test it against the underside of your wrist. It should feel warm but never hot enough to cause discomfort.

If the mixture has gone too cold by the time it has swelled to its final consistency, you can rest the cup in a bowl of boiling water for a short time. Formula that has been reheated must be thoroughly stirred and the temperature retested before feeding to ensure that there are no hot spots and that the formula has not been overcooked. Heating in a microwave can cause crop burn due to the formation of 'super-hot' spots in the formula.

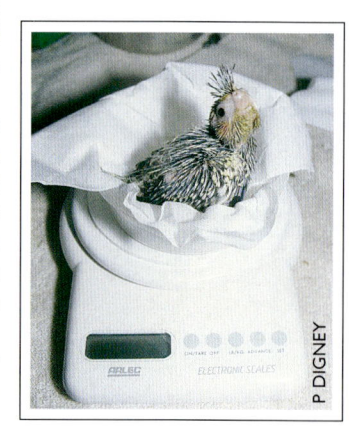

Daily weighing of chicks is essential.

Feeding Instruments

All feeding implements and mixing cups need to be sterilised prior to use. Milton™ human infant sterilising solution and other suitable veterinary disinfectants are readily available. Submerge feeding cups and spoons in the solution at the recommended strength. At the start of each day, also submerge a water jug in the solution for the recommended time. Rinse the jug a few times with boiling water before filling it with boiled water that is left to cool.

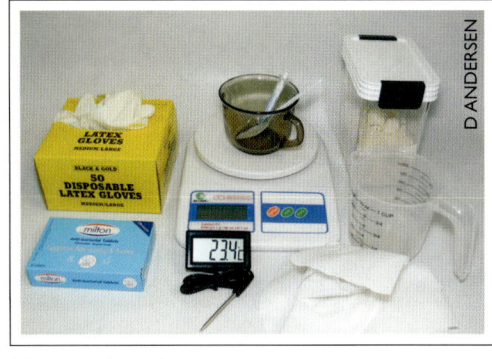

Various handrearing requirements.

Prior to each feed, rinse the mixing cup and spoons in boiling water and allow them to cool a little before mixing the food. This is done to remove the antibacterial solution from the feeding utensils to prevent the chicks from ingesting some of the solution. Although this is not required for human babies I feel that it is a sound practice with baby birds as antibacterial solution in their crops could predispose chicks to fungal infections. Digestion requires the presence of friendly bacteria in the crop. Destruction of these bacteria can upset the delicate balance in the crop and result in the growth of yeast organisms.

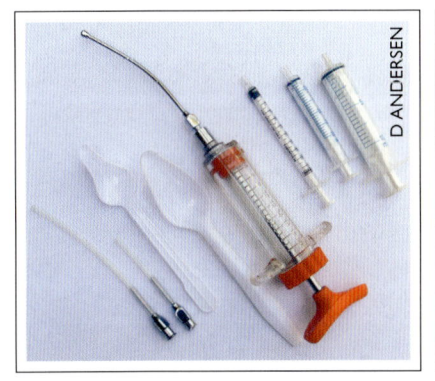

Handfeeding can be done via spoon, crop tube or syringe.

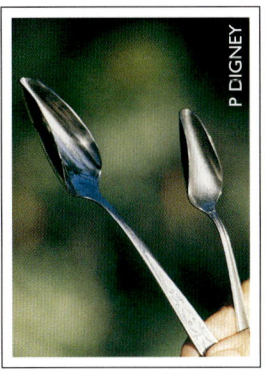

Simply bend up the sides of a teaspoon for spoon-feeding.

Spoon-feeding can be messy. Clean the chick's face with tissues after each feed.

Feeding Methods

There are many different methods of handfeeding chicks.

I spoon-fed for many years, using disposable plastic spoons for both mixing and feeding. These spoons can be easily bent into the correct shape for feeding by submerging them in boiling water for a short time and pinching the sides together when the plastic has softened. They do not retain heat like metal spoons and are therefore unlikely to burn a chick's mouth. They are also cheap and can be readily discarded. Spoon-feeding is the safest method for inexperienced people to use and the most natural way to feed. It triggers a

Spoon-feeding is a practical and simple method of feeding.

feeding response and is similar to normal parent feeding. There are fewer weaning difficulties with spoon-fed chicks than with tube-fed chicks, and less likelihood of the young aspirating the formula.

If you are rearing a number of chicks, spoon-feeding can be impractical. The alternative is syringe-feeding. Unlike tube-feeding, which largely bypasses the bird's tastebuds, syringe-feeding still triggers a natural feeding response and allows the bird to experience taste. This is beneficial when it comes to weaning. Another benefit of syringe-feeding is that you can control the amount of food that a bird receives at each feed. In addition, syringe-fed chicks generally remain much cleaner as there is less spillage. The syringe is placed in the side of the beak and touching the corner of the mouth. This generally triggers an automatic feeding response. Pushing the plunger down should not commence until the feeding response has started. This response opens the oesophagus and allows food to pass easily into the crop.

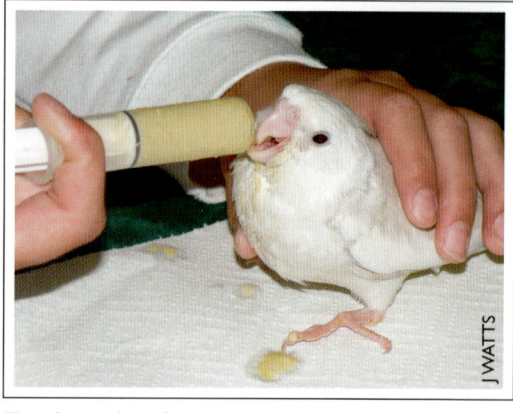

Feeding this five-week-old chick by syringe is also a simple process, as shown by this nine-year-old bird keeper.

It is easy to aspirate a chick that is resisting being fed. Even with a feeding response, you need to have a steady hand and a syringe that depresses smoothly. Lumpy, gritty formulas are not suitable for syringe-feeding. Blockages can cause resistance that allows pressure to build up. When the blockage passes, the formula can then squirt suddenly into the bird's mouth and may cause aspiration.

I do not recommend tube-feeding for a number of reasons. As mentioned earlier, it does not allow the chick to experience taste. Unless you use a different needle for each chick or allow time to disinfect the needle between chicks properly, it can rapidly spread infections from one chick to another. The incubation time for some viruses would allow many chicks to become infected before the first chick started showing signs of the disease.

You can also miss early signs of illness by tube-feeding. Often a chick that is becoming ill will beg for food but will only take a small amount before refusing food. If you are tube-feeding the food is delivered rapidly in one dose. Unless the chick starts to regurgitate you may be unaware that it is getting ill. Chicks that are fed with a tube may also receive more food than they require as they approach fledging, a time when they would normally begin to refuse food in order to lose weight in preparation for flying.

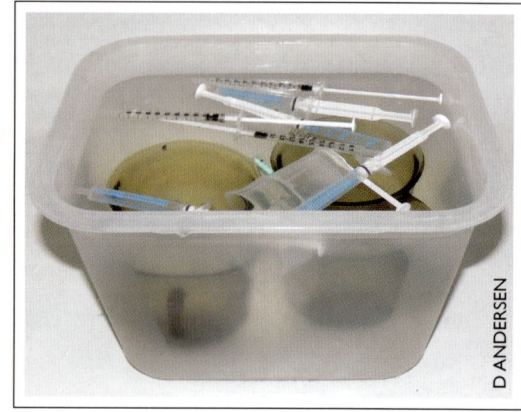

Hygiene is vital, with disinfection of all implements in Milton™ solution required between feeds.

Overfeeding at this age can result in regurgitation, simply because the chicks are too full. Cockatiels should never regurgitate as youngsters. If you see this happen, immediately suspect a health problem and seek veterinary help.

If not using a bib, after each feed the chick should be cleaned of any residue food.

Feeding Regime

Chicks are fed every 1½–2 hours for the first three days except through the night, when they are fed at midnight, 3.00am and 6.00am. The two-hour schedule then continues. On the fourth night, the 3.00am feed is excluded. As the food thickens and the crop increases in capacity, feedings can be spread further apart. Chicks should be fed when at least three-quarters of the crop contents have been digested.

Regularly scheduled feeds should continue throughout the day until about 11.00pm. However, chicks should not be fed in the morning until the crop has emptied from the night before. If this takes too long, the chick may draw the remaining moisture out of the crop, making the contents thicken to where they cannot be digested. In this case it is better to feed a couple of diluted feeds and gently massage the crop contents so that the compacted food can break up and digest.

Probotic™, containing eight strains of friendly bacteria, is recommended for young chicks and birds that are ill or have slow-moving crops. Spark™, on right, is an excellent rehydration supplement.

If the crop has become pendulous, ie overstretched, usually as a result of overfeeding, it will not empty and the chick will simply starve while you wait. In these circumstances it is better to continue feeding while massaging the contents of the crop gently so that some of the old food digests and passes out through the digestive system with the new. Nilstat™ should also be used for this situation. Old, undigested food lying in the bottom of a crop is a perfect breeding environment for the fungus *Candida albicans*, the cause of thrush. Forcibly emptying the crop can be beneficial in removing the old food that has gone sour but it is a very stressful experience that can make a weak chick even weaker. Therefore, it should not be attempted if you do not have the skill. As mentioned earlier, suspected bacterial infections are best treated by an avian veterinarian who will administer the correct antibiotic at the correct dose.

As chicks approach fledging they usually require only three meals a day, showing least interest in the morning meal and most in the evening meal. Some will only take a reasonable feed in the evening. This is not a cause for concern as long as they appear bright and well. After fledging, their appetite usually returns to normal.

Chicks that are removed from the nest by choice, not by necessity, are best taken at three weeks of age. At this time they will generally adapt quickly to handfeeding and are not as delicate as younger chicks.

Chicks that are closer to fledging can be difficult to feed as they are not as hungry and can be very stubborn. If they open their mouth to hiss at you, you can quickly spoon or syringe a little food into their mouth. If they just sit and glare at you, have patience. Some chicks can go up to two and a half days without a reasonable feed. In situations such as this, it can be helpful to sit quietly with the chick, stroking it until it loses its fear. Placing it with younger chicks that are feeding well can also often encourage a feeding response. There is little choice with chicks that have already fledged other than to feed them with a crop needle.

Monitoring Chick Development
Weight Charts

Weighing chicks—and comparing the results against an average weight chart—is an ideal way to monitor a chick's development. Often signs of ill health are indicated by weight loss days before a chick begins to show any other symptoms. Chicks should be weighed each morning at the same time when their crop is empty.

Following is an average weight chart for a chick that was handreared from day one.

Cockatiel Handrearing Weight Chart

Day	Weight in Grams
0–hatch	3
1	3
2	4
3	5
4	6
5	8
6	10
7	11
8	14
9	15
10	18
11	21
12	23
13	26
14	29
15	33
16	38
17	44
18	50
19	57
20	63
21	68
22	74
23	79
24	86
25	91
26	95
27	97
28	96 (peak weight)

Above: Normal (with yellow down) and Whiteface chicks at seven days of age.
Below: Chicks at three weeks of age.

Although weight charts can be beneficial, they are a guide only. Chicks will grow at different rates for many reasons, eg choice of formula, hatch size, feeding technique, health issues such as hydration and, of course, genetics. It is important not to overstretch the crop trying to get a chick to digest more food in order to match the progress of another chick based on a weight chart.

Fully feathered young Cockatiel.

Chicks that grow at a slower rate can often end up just as large as a chick that has gained weight more rapidly. Provided that a chick is not being underfed or reared on a diet that is nutritionally deficient, most will reach the size that they were genetically determined to reach.

Body Condition

Although weight charts are a good way of monitoring chick development, a good handfeeder should also be able to assess a chick's progress through its body condition. Regardless of how quickly or slowly a chick gains weight, it should always have a full rounded breast and the keel bone should not protrude sharply from its chest. The chick should have an overall plump appearance and a healthy pink tone to its skin. A chick that becomes pale is a cause for concern. This can be indicative of a viral or bacterial infection and will generally be accompanied by poor weight gain, listlessness, poor feeding response and abnormal droppings.

Chicks that are strong when they hatch will be sitting upright begging for food within hours. A healthy chick should continue to wake at regular intervals and beg for food until it approaches fledging when it becomes even more active, even though it may lose its appetite.

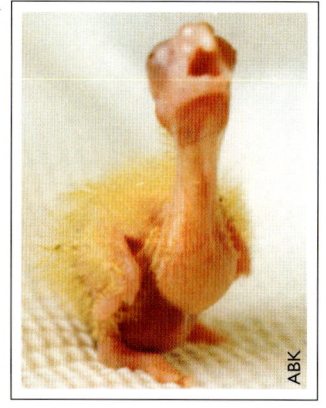

Strong chicks sit upright while begging for food.

It is important to avoid dehydration at all stages of the chick's development. Chicks that are brooded at too high a temperature can become dehydrated as can chicks that are kept too cool—due to the stress that the chilling causes. Chicks should not have dry, scaly or reddish skin. The tightness of skin in dehydrated chicks gives them a bony appearance. Underfed chicks can also become dehydrated as their fluid intake is often also inadequate.

The bone structure of a chick should develop in proportion to the rest of the chick. Formulas that are too high in fat can cause the chicks to become overweight. The excessive weight can place abnormal stress on the growing bone structure. As with chicks in the nest, you should always monitor leg development to avoid conditions such as splayed legs.

Hatchlings begin to open their eyes from seven days of age. The eyes will take several days to open completely but should remain clear and free of any discharge once they have fully opened. (See also *Feather Development* on page 52.)

Weaning

In an effort to force chicks to wean at a particular time, it is common amongst handfeeders to reduce the number of feeds. It has been my experience that Cockatiels can be 'forced' to do very little at all. The reduction in food seldom helps chicks wean and more frequently results in hungry, insecure and distressed chicks that become dangerously thin and obsessed with being fed. Once they begin to behave in this manner, weaning becomes a very difficult and lengthy procedure.

Handraised fledged young are provided the opportunity to investigate food items around their playgym.

A variety of soft vegetables assists young in the weaning process and accustoms them to a healthy range of foods.

The process of weaning stems from the natural curiosity of the chicks picking up, nibbling and exploring items in their environment. In happy, contented chicks, this process commences in the nestbox. After fledging the chicks learn from parents and companions that have already weaned.

Branches and seeding grasses are most readily accepted for nibbling. A bowl of dry Budgerigar seed mix and fresh water should always be available from the time the chicks fledge. If you feed pellets, wean the birds onto seed first and then make the conversion. This is advised because it cannot be guaranteed that the people who purchase your birds will feed pellets. Therefore, the youngsters must know how to crack and eat seed. Chicks weaned only on pellets (which have a flavour and texture similar to the handraising formula) often fail to learn how to crack and eat seed. It took me six months to get one young bird to accept seed after being weaned onto pellets. However, the conversion process from seed to pellets in young birds that are eating lots of new things has never proved difficult for any of my birds. In older birds, however, it may be difficult and in some cases not advised if they do not appear to accept pellets after some time.

Prior to fledging, handreared chicks always investigate seed and other bits scattered on the floor. Soft vegetables, eg corn on the cob, peas and apple also help the weaning process, as they provide the chicks with something to nibble on. In this way they experience taste and swallowing follows shortly afterwards. Chicks that suddenly have their food intake reduced, before having learnt to swallow, can become panicky and insecure and cease to explore their environment, choosing instead to sit and listen for the sound of your approach, begging constantly for food.

Chicks will frequently wean much more rapidly onto softfoods such as sprouts. However, they should also be provided with other items to nibble at, eg dry seed and vegetables. In addition, eating only sprouts can predispose chicks to future fungal infections.

Handfeeding chicks can be a tedious process and the desire to wean the chicks too soon for our own convenience becomes very strong. To alleviate this problem with chicks that are proving difficult to wean, instead of starving them, teach chicks of 6–7 weeks of age to eat from a bowl. This is easily done by lowering a spoon or syringe of handrearing mix into the bowl of food when the chicks are begging. You can also encourage the chicks by feeding mouthfuls of the formula from just above the level of the bowl, causing them to lower their heads and grab small mouthfuls of food from the bowl. Once they understand where the food is, leave the spoon in the bowl and quickly leave the room. The handraising mix used in this weaning process should be quite thick and similar to the consistency of porridge.

Chicks that are old enough usually learn to feed from the bowl after two or three feeds. When you return to the room always check their crops to make sure that they have been adequately filled. The chicks can be fed up to four times a day with minimal inconvenience when they learn to feed this way. They can eat their fill at their leisure and maintain a happy contented disposition. With this method chicks quickly learn to associate eating food with bending down and picking up small amounts of food rather than bending their heads back and receiving food from a spoon, tube or syringe.

Weaning Cage

Allow chicks to fledge in their own time by placing the plastic tub in which they have been brooded into a weaning cage. They will gradually climb out of the tub when they are ready. By reaching into the cage, several chicks can be fed at once. This saves time and keeps them interested because they also compete with their perch mates for the food. In addition, older chicks are easily distracted and too busy to eat their fill, preferring instead to investigate their surroundings. Feeding them inside the cage also prevents the dangerous thrill of flying around the room and into harmful objects or windows.

An excellent set-up of weaning cages, including an enclosed plastic 'drawer' for housing chicks from 3½ weeks to fledging age. No extra heating is required. Fledglings are then placed in weaning cages and allowed out on the playgym.

As chicks approach fledging, solid food can be introduced to the brooding environment.

A weaning cage constructed of 12.5mm x 12.5mm wire mesh and measuring 80cm long x 60cm square is suitable. The base can be a tray which will catch droppings and food items that fall through the wire base.

Chicks lose some weight during the weaning process. Therefore, as an extra support in cooler weather, juveniles should be given access to a heat source at night until they have weaned and gained considerable weight. Locate a 60-watt lamp near the weaning cage to prevent the chicks from coming into direct contact with the heat lamp. A perch can be positioned such that the chicks can move towards or away from the heat source if desired.

For more detailed information on artificial incubation or handrearing consult a more detailed text on the procedures. Highly recommended is **A Guide to Incubation and Handraising Parrots** by Phil Digney, published by **ABK Publications**.

In cooler climes a 60-watt lamp provides a heat source for weaning birds.

BASIC FIRST AID FOR BIRDS

Birds, like children, always seem to be getting themselves into trouble. Knowing some basic bird first aid may mean the difference between life and death for your pet.

Your bird's **first aid kit** should include:

- Styptic powder, Condy's crystals or a ferric chloride stick to stop any bleeding from toenails;
- Electrolytes, such as Spark™ or Electrovet™, to add to the water when the bird is ill or stressed, or when stress is likely to occur;

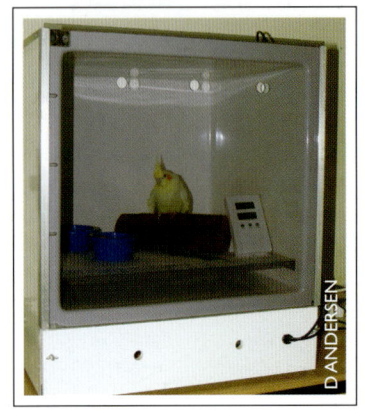

Above: This brooder provides warmth and isolation for sick birds.

Below: A heat lamp can be positioned on the top or clipped to the side of a cage. Note the distance between the bulb and the cage.

- Paper towels;
- Eye droppers for emergency handfeeding;
- Your avian veterinarian's clinic and emergency telephone numbers.

Hospital Cage Accessories

You should always have a hospital cage at the ready. (This could even be an empty aquarium-type tank with a mesh cover.)

Hospital cage accessories include:
- A cage cover;
- Clean towels to handle your bird and, if necessary, to cover the bottom of the hospital cage (a baby blanket could be used for this);
- A heat source such as a heat lamp, a heat pad or a reading light (with a 60-watt pearl or coloured bulb);
- A thermometer to check heat levels;
- A humidity reader to check humidity levels.

Hospital Cage Set-up

The hospital cage should be almost completely covered. Warmth should be provided via a heat lamp or heat pad set at 27–33°C. If using a heat lamp, arrange it so that the bird can get as close to or as far away from it as it wants. And ensure that the cage cover cannot touch the heat lamp—you do not want to start a fire. Monitor the heat and humidity levels and, of course, your bird. If the bird is panting and/or holding its wings away from its body, decrease the heat. Food and water should be easily accessible. Ensure that the bird is eating and drinking. Offer the bird its favourite foods (this is not a time to be miserly with treats). If you do not feel that it is consuming enough food and water, handfeed your pet with an eye dropper or spoon. If you do not have any handfeeding formula, use warmed human baby food, eg beef and vegetables, or glucose and water. Be aware that birds dehydrate rapidly and therefore their weight should be monitored carefully. If your bird is losing weight or does not improve in a short period of time, call your veterinarian immediately.

Common Injuries

The most common (and easily treated) injuries that you will have to deal with are a broken blood feather or a clipped toenail vein. As birds succumb quickly to blood loss, it is important to act promptly. After washing the wound with cool water, apply pressure to the injury. In the case of a toenail, press styptic powder or Condy's crystals into the wound. Keep the pressure constant until the bleeding stops. If the bleeding from a broken blood feather persists, use needle-nosed pliers to pull the feather out, and then apply pressure to the follicle until the bleeding stops. If bleeding from a toenail or blood feather cannot be stemmed, take your bird to an avian veterinarian.

When you suspect that your bird is not feeling 100%, attend to it. At our aviary, should a bird appear even slightly fluffed up or have a hint of a red nostril (which usually turns out to be from pellet dipping!), give the bird warmth, quiet and privacy. In most cases, the bird returns to normal within a couple of hours. Probably nothing was wrong in the first place, but it never pays to take chances—if in doubt, give your bird heat!

COMPANION BIRDS

Whiteface Grey Pearl Cockatiel.

The Cockatiel is one of the most commonly kept pet birds. Its small stature and relatively quiet voice, its ability to talk and the fact that many Cockatiels seem to delight in the company of their human owners, have made them an extremely popular cage bird.

Whether you choose to purchase a single bird or a pair of birds you will quickly realise that these delightful parrots find great enjoyment in sharing their lives with you. They derive genuine pleasure from human affection and will reward you with years of companionship and entertainment. In return they deserve the best care and attention that you can provide.

Choosing a Pet Bird

Cockatiel personalities vary greatly and not all birds make good pets. A breeder or specialist pet shop should be able to supply you with a pet that comes from a particularly gregarious and friendly family. These birds, when handreared, often remain tame even in aviary situations with mates that are not tame. In the past, people would simply purchase a wild or aviary-bred, parent-raised Cockatiel and gradually tame the bird by repeated exposure to the hand of their owner. This requires a great deal of patience and, depending on the owner and the approach that is used, may be totally unsuccessful. It is also a very stressful experience for the bird and can even be damaging to the health of a bird who finds this sort of technique terrifying. There are many breeders and pet stores that offer handreared birds for sale. Considering the amount of work involved in handrearing a Cockatiel, it is well worth paying more for a handreared bird that does not require a major effort to tame.

The small stature and quiet disposition of Cockatiels make them ideal pets for young bird lovers.

As far as which sex makes a better pet, both have attributes that make them ideal. Cocks have a tendency to be very active, vocal and entertaining, although some are not happy to be stroked and scratched, preferring instead to ride around on your head or shoulder. Hens, on the other hand, are very sweet and affectionate and will often gently insist that you scratch their heads for hours on end. They are frequently content to nuzzle into your neck and sit quietly for long periods.

When selecting a pet bird the same principles apply as discussed in the section *Acquiring Stock* on page 23. Choose a bright, active, healthy-looking bird with a good body weight. Juvenile birds are a similar weight to adults, perhaps a little heavier before weaning. A healthy body weight is 80–100 grams.

Handraised Cockatiels may not be in perfect feather condition because the feathers around their beaks and on their chests may be matted with handrearing formula. However, a good handrearer will keep this to a minimum. After weaning, the birds will soon preen their feathers and look presentable. Because Cockatiels often climb around the front of the cage trying to get food and attention they may have a few broken tail feathers. Although these features are not always indicative of an unhealthy bird, an effort should be made not to make too many excuses for a bird in poor condition. It should not be dirty and underweight as a result of being handreared.

It is not necessary for a new owner to complete the handrearing process in order to make the bird bond to them. Handrearing can be a tricky business and is best left to people with the required expertise. If the bird has been well handled during the rearing process it will adapt quickly to its new owner, provided that you adopt a 'slow and steady' approach when you bring your new bird home.

It is important to purchase fully weaned and independent young.

Young Cockatiels lose weight during the weaning process and can become ill if placed under the stress of being transported to a new home, pet shop or bird dealer. It is important to purchase a bird that is fully weaned—and is healthy and eating well on its own.

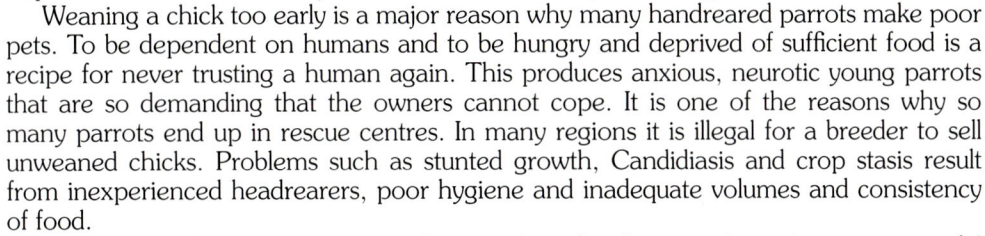

Weaning a chick too early is a major reason why many handreared parrots make poor pets. To be dependent on humans and to be hungry and deprived of sufficient food is a recipe for never trusting a human again. This produces anxious, neurotic young parrots that are so demanding that the owners cannot cope. It is one of the reasons why so many parrots end up in rescue centres. In many regions it is illegal for a breeder to sell unweaned chicks. Problems such as stunted growth, Candidiasis and crop stasis result from inexperienced headrearers, poor hygiene and inadequate volumes and consistency of food.

Being moved to strange surroundings with unfamiliar people is the most stressful time of a young Cockatiel's life. Even if the juvenile appears to be weaned, the insecurity brought on by a change of environment often results in weaning regression, ie the bird begs incessantly for food and becomes overly dependent on the new owner (McKendry 2006). It might pick up an item of food, then drop it again, because it needs something warm and soft to eat and may even nip the owner who does not realise that the bird is hungry. This is not the way to gain the love and trust of an innocent young bird. If such a situation occurs, handfeeding—offering an abundance of palatable foods—must be carried out in order to prevent the bird from becoming dehydrated and anxious.

Avoid startling your new pet with such actions as wiggling your finger which can invite a bite. It is better to offer a food treat slowly. Patience is the key.

One Pet Cockatiel or Two?

It is my personal opinion that a Cockatiel will be more content if kept in the company of another bird, preferably of the same species.

Many people believe that a pet bird needs to be kept solitary in order to remain tame and fill the role of companion to their human owner. Because Cockatiels are a flock species, it is true that they will become devoted to their owners. In the absence of another bird, humans become their flock and their desire to bond with a mate is redirected towards their human owner. The problem is that humans are often fickle creatures and their circumstances may change. They may no longer be able to give their bird the attention it requires. Children often give their bird a great deal of attention when it first arrives but then tire of their new companion. The bird will then spend more and more time confined to a cage. Behavioural problems such as screaming for attention and feather plucking may soon arise, as the bird becomes bored and lonely. Many people forget how long their birds might live. At the time of writing my oldest Cockatiel is 18 years of age and still going strong.

It has been my experience that two Cockatiels can be kept just as easily as one and these birds can entertain each other while their owner is at work or when unable to spend time with them. Provided that both birds are tame and they are invited to spend time with their owners, there is little chance that they will revert to rejecting human contact.

Being a flock species, Cockatiels rely on other members to warn them of potential dangers. Flocks will take to the air instantaneously at the first sign of danger. Solitary Cockatiels can therefore be more nervous as they have no-one else to rely on for their safety from potential threats. With the introduction of a second tame bird, people who have acquired a handraised bird from me as a companion for an existing pet, have told me that the personality of their first bird had blossomed. This may have resulted from a boost in the bird's self-confidence due to the fact that it is no longer alone. None have reported deterioration in their relationship with their first bird. Being a flock species means just that. While their mate may be another bird, it does not prevent them from wanting to interact and spend time with other 'flock' members who are always good for a scratch!

Cockatiels are a flock species and will generally benefit by being kept with one or more mates.

A companion bird for your Cockatiel need not be of the opposite sex and, in my opinion, a true pair should be avoided if you do not wish to breed from them. The cock may stimulate the hen to breed, resulting in excessive egg laying.

There is also the belief that a bird will not learn to talk in the presence of other birds. The ability to mimic and the desire to do so varies from bird to bird, with cocks usually more inclined to learn words. I do not teach my birds to talk but one of my cocks taught himself to 'wolf whistle' and to make a noise resembling 'Donald Duck'. Being a very vocal cock, he managed to teach all my other cocks and some of my hens to make the same noise. I am quite sure that if I encouraged him, he would learn to talk and probably teach many birds in my flock to say the same words.

If you purchase a bird that is not tame as a companion for an existing pet bird, I recommend taming the second bird before introducing the two to each other. Because they are a flock species, the fear that the new bird experiences in its new environment may be felt by your existing bird and may alarm it.

Taksan, at 28 years of age, could be the oldest recorded, living companion Cockatiel.

Veterinary Check

Birds should be checked for internal and external parasites before being offered for sale. It is advisable to request a health guarantee for a period of at least four weeks after purchase. In the event that a bird dies during this period, an autopsy must be performed to determine the cause of death. This will isolate any health or environmental problems as the cause.

Because Cockatiels are relatively inexpensive, a breeder or pet shop may not be willing to offer to supply the bird with a veterinary health certificate. If you cannot obtain a health guarantee from the seller, then it may be worth asking an avian veterinarian to check the bird as soon as possible after purchase. This will enable you to identify immediately any potential health issues that may cause problems for the bird. Without going to too much expense, a good avian veterinarian can identify many parasitic problems, perform some basic checks to screen the bird for bacterial and fungal infections and also give you an informed opinion on the general health of the bird. Avian medicine is a highly specialised field so it is also wise to identify your nearest avian veterinarian in case of a future emergency. (See *Health and Disease* on page 178.)

Preparing a Home for Your Pet Cockatiel

Over the past few years a much wider range of attractive cages have become available to the pet market. Unfortunately, the smaller the cage, the more likely it is to be unsuitable for its occupants. This is because the people who design many of these birdcages seem to have little understanding or knowledge of the requirements of the eventual inhabitants.

Each year small cages become taller, narrower and less suitable, although they may seem more attractive from a decorative point of view. Birds fly in a more or less horizontal fashion. Cockatiels fly rapidly up and down only when responding to a sudden fright. Therefore, cages that are longer and wider than their height are much more suitable for pet birds than tall, narrow styles. Cockatiels that are tame and confiding to their owners can usually be released to fly at liberty in an area of the house. If they are allowed to explore and exercise freely for a period each day, the size of the cage is less critical to their health. For a pet Cockatiel allowed regular exercise, I recommend a minimum cage size of 60cm long x 45cm wide x 45cm high.

In order to provide enough length for two birds, you should purchase a cage designed for larger species. A small cockatoo cage, measuring approximately 90cm long x 60cm wide x 60cm high, would allow two pet Cockatiels to be housed comfortably together.

Although larger cages are always better if you have the space, they do not always have the most suitably spaced bars for the welfare of your bird. Care must be taken not to leave cages of this size unattended outdoors. Cockatiels can easily put their heads through the bars and are in danger of losing them to owls, hawks and cats. Cages with horizontal bars are more suitable for Cockatiels to climb around on than the usual vertical bars used in Budgerigar and canary cages.

It is possible to make your own cage from aviary weldmesh, using 'J' clips to fasten the joins. Ideally these cages should measure 90cm long x 60cm wide x 60cm high and have a feeding veranda at one end. These cage dimensions, suitable for a pair of Cockatiels, will promote the good health of your birds.

Cages that are longer or wider than their height are much more suitable for pet birds than tall, narrow styles.

Such a cage is economical to make, easy to maintain and light in weight, making it ideal to carry outside so that the birds can enjoy some fresh air and sunlight. The weldmesh is much more suitable for climbing and can be painted black with water-based paint to make the birds easier to see and the cage more attractive. A large tray placed under the cage will provide a catchment area and keep the surroundings clean.

Perches

It is advisable to provide natural perches rather than the dowel perches that usually come with a new cage. The varying natural dimensions of branches create exercise for the birds' feet and stripping bark seems to be an activity that keeps many Cockatiels occupied for hours.

Some new perches that have recently become commercially available are smooth on the top where the birds' feet rest but have an abrasive pumice coating on the sides that keep their nails trim. These perches have varying dimensions along their length, which again helps to exercise the birds' feet. Sandpaper perches are not good for birds as their abrasive nature can damage the soles of their feet.

Natural perches of varying dimensions provide exercise for the feet.

Food and Water Dishes

Food and water dishes should be placed where they can be easily accessed for cleaning and refilling. However, they should be placed away from the perches so that they do not become fouled. Food dishes should also be located away from water dishes to avoid spoilage of the water. Many Cockatiels have a cage set-up that provides little motivation to move around. Locating food and water dishes in separate areas around the cage is the best approach to changing this. It is also an essential component of effective environmental enrichment. Creative food-dish placement will encourage your Cockatiel to move around and exercise.

Food and water dishes should be placed apart to prevent spoilage and provide exercise. The wire cage floor allows spilt food and droppings to fall through to a pull-out tray below.

The Cage Floor

Many cages have wire bottoms that allow pellet crumbs, seed husks and other spilt food items and droppings to fall through onto a tray below. This prevents birds from coming into contact with old and fouled food items like fruit and vegetables and restricts the bird's access to its own droppings. This can be beneficial from a health perspective. However, because Cockatiels are terrestrial foragers, it is preferable for them to have access to an enclosure floor. This will provide opportunities for natural foraging behaviours and reduce the development of stereotypical behaviours. Provision of a special 'foraging tray' on a regular basis will compensate for the lack of ground-foraging motivation in a captive environment, particularly if the substrate is wire.

If the cage you choose has a solid floor it is important to maintain the hygiene of the cage. The floor can be covered with newspaper. This must be replaced regularly. And a special effort needs to be made to remove old food before it gets a chance to spoil.

Cage Location

Because pet Cockatiels are usually housed inside, their owners may not consider aspects such as draught or cold. Many behavioural abnormalities and health problems are caused by unsuitable environmental conditions. For instance, many Cockatiels are housed in the living areas of the home. When the human inhabitants require heat the room is heated for their comfort. Then the humans turn the heat off and crawl under blankets for the night, leaving their pet Cockatiel to experience a rapid drop in temperature. Cockatiels moult and breed in response to the change of season and the length of daylight hours. A pet Cockatiel experiencing an unnatural range of temperatures and daylight hours will often develop feather problems, as well as behavioural problems such as feather plucking as a result of this disruption. A simple solution is to keep your birds in a room that is not artificially heated during the day and move the birds to an unheated room when you turn the heat on in the evening.

A view from the window may seem to provide entertainment. However, birds such as this Green-cheeked Amazon and this Cockatiel, can be easily frightened by cats, dogs or wild birds. The location may also become a hot spot on sunny days if the sunlight is not filtered. The cage should be located further back from the window.

It is also helpful to cover the cage in the evening if the room is draughty. When it gets dark outside, lights should be turned off or the cage covered. Cockatiels housed outside naturally roost from the onset of evening until dawn. Many people do not realise that their Cockatiel is not genetically programmed to be up and active at midnight!

All birds enjoy basking in the sun and this is especially true in winter. Sunlight is important for the production of vitamin D_3 and the absorption of calcium. While it is relatively safe and very beneficial to hang your pet's cage in the sun in winter, the birds can die rapidly if left for too long in full summer sun. Ideally, the cage should be located where it receives morning sun only and even then it should always have some area that is shaded inside the cage so that the bird can move out of the direct sunlight if it wants to.

Yellow and green Budgerigars have been used as bait to trap and relocate troublesome raptors, because hawks and other predator birds are able to spot them from a distance of up to 5km away. Your pet Cockatiel of any colour would be just as effective as bait! Consider the hanging position of your cage carefully and try to provide some visual cover. If you are not going to be within hearing distance, move the cage back inside. The distress call of a Cockatiel is easy to recognise and quick action from the owner can usually rescue a pet from a potential predator.

Some recent studies have suggested that locating birds near a doorway can increase the likelihood of stress-related behavioural issues such as feather plucking. In such locations the bird may be easily and frequently startled by the sudden opening and closing of the door as well as the increased traffic infringing on the bird's personal space.

The First Days at Home

Cockatiels that have been handreared will be tame and confiding to the person who raised them but can still be wary of strangers. There are some important steps to take when first introducing a young bird to its new home.

Your new pet bird should remain confined to its cage for at least a day until it becomes accustomed to the voices and appearance of its new owners and the sounds of its new environment, especially any new animals. If there are several children in the family, it is preferable that only one adult and one child at a time work with the bird. A room crowded with unfamiliar people can panic a bird.

The bird should be kept in a smallish room initially. During this time you should focus on building trust, familiarity and predictability with your pet bird still in its cage—especially if it is not handtame. This development may take time. It is important to provide your bird with a secure environment, spending regular, short periods gently interacting and observing the bird's reaction to better understand its predictable behaviour. When you are confident that the bird will accept food treats from the hand while it is still in the cage, then it might be time to encourage it to hop onto your hand and be lifted out.

A bird that is allowed to come out and

Allow out-of-cage time for your pet to explore its environment.

Family members are encouraged to spend time to develop a relationship with their pet Cockatiel.

explore its environment without fear will quickly become accustomed to its surroundings and its new owners. The smallish room prevents the bird from flying too fast and possibly injuring itself or becoming alarmed by unfamiliar objects in the room. The focus must be on developing trust by taking small steps.

If the bird initially flies away from you and lands on the curtain rail, allow it to settle before quietly approaching it and encouraging it onto your hand. Provide eucalyptus or seeding grasses for the bird to chew on while it is in your lap or on your hand. As well as being a treat, it gives a nervous bird something to do and tends to have a calming effect. As the bird becomes accustomed to you and its new surroundings, it will become increasingly bold and can be allowed to explore larger areas in the house.

Diet

Diet has been discussed in depth elsewhere in the book. However, it is worth mentioning a couple of issues that are especially relevant to pet Cockatiels. Initially, provide food items to which the bird has become accustomed and gradually make any dietary changes, such as conversion to pellets and other nutritional food items. Apart from the fact that a balanced and varied diet is nutritionally better for your bird, it also allows you to utilise food treats as rewards for desired behaviours such as stepping up onto your hand, flying down to you when called, returning to its cage by choice or even going into a carry box when the bird needs to go to the veterinary clinic.

These treats should be food items that you know your Cockatiel likes. Treats used to enrich a bird's diet or as training aids should be used as such and should not be given in quantities that prevent a bird from eating its normal diet.

Some people allow their pet Cockatiel to eat from their dinner plates. While some items on your plate might be safe and healthy for the bird to eat, other items such as potato chips and other fried foods are not. Sugary drinks, coffee, alcohol, chocolate and avocado are also unsuitable for birds and may be harmful to their long-term health.

Cockatiels that are used to eating a varied diet, including fruits and vegetables, should be monitored to ensure that they are not supplementing their diet on their own by nibbling on potentially poisonous house plants while at liberty in your home. (See *Feeding* on page 35.)

Enrichment

While the large cockatoos may be commonly considered to be more intelligent than their much smaller relatives, Cockatiels will still become easily bored if left in a small cage with little to keep them occupied for hours on end. This boredom leads to behavioural problems which are difficult to reverse once they have begun. Therefore, Cockatiels should be exposed from an early age to a variety of enriching experiences.

Toys

The provision of toys not only relieves boredom and encourages independent play but also enriches your bird's life. Thankfully a variety of parrot toys are available on the market today. Try to purchase toys that encourage natural foraging behaviours; toys that amuse and promote physical dexterity; and toys

A variety of toys will provide enrichment for your pet.

that provide an outlet for their chewing and/or preening behaviours. Note: mirrors seem more of a torment for Cockatiels than an enrichment item so I have always avoided providing them. A real Cockatiel makes a much better companion for your bird than a mirror image!

Most parrots need one or two toys in their cage. Rotating them on a regular basis, eg every week, will increase their enrichment potential—and your pet bird will not become bored with them.

Most parrot toys are safe, while others present various hazards that are sometimes difficult to identify even to an experienced eye. Be aware of items that fray easily or that your bird could get caught up in. It is important to monitor the state of all toys regularly and discard items that may have become a potential hazard.

Cockatiels should be provided with a variety of natural foraging materials for enrichment.

A variety of toys will keep your pet entertained.

Natural Branches

You need not spend a fortune on purchased toys. Like children, Cockatiels can tire of the same toys whereas fresh native branches and seeding grasses, items that are part of their natural foraging and nesting behaviour, will often keep them busy for far longer than man-made toys.

For those who are lucky enough to live in Australia, the Cockatiel's native habitat, enrichment items can usually be obtained within a short walk from the front door. Most native Australian plants have buds and seed pods, bark and leaves that are safe for your bird and will keep it busy chewing for long periods. This often has the added advantage of filling your house with fragrant odours such as eucalyptus.

Flight

Allowing your pet Cockatiel to fly enriches its life and gives it a great deal of freedom and independence when out of the cage. A bird that can fly is a pleasure to watch and if your bird flies down to land on your head or shoulder you can be certain that the bird wishes to spend time with you. I always feel in some way privileged when this happens.

Unfortunately there are also dangers associated with flighted birds. Birds often do not see the glass in windows and doors and can crash into them with enough force to cause serious injury and even death. Even if they are used to a room and are aware of the glass, they can often forget when they are startled. As mentioned earlier, it is best to draw the curtains if your bird is out of its cage. Owners, especially children, can sometimes forget that their bird is on their shoulder and walk outside. The bird, alarmed at the sudden change from its familiar environment, will often fly off and become lost. Many notices in the 'Lost and Found' column of the newspaper are testimony to this.

It is better to restrict birds to the innermost rooms of the house rather than giving them free range of the whole house. Flighted birds are not always selective in their landing places so it is better to return your bird to its cage if the stove is on or the sink is full of water. Ceiling fans are also a major threat to flighted birds.

Basic Behaviour Training

Engaging your pet bird in basic behaviour training will teach the bird what is expected and assist in the bird's understanding of what behaviour is being encouraged and what your actions are likely to be. This will also provide you with the opportunity to interact positively with your bird. Start with commands such as 'step up' or 'hop down'. As the bird learns to respond to these requests, he should be rewarded, if possible, with a treat. This is the basis of a good relationship with your bird.

When returning the bird to its cage, supplying a new enrichment item within the cage when you request the bird to 'hop down' makes returning to the cage a positive experience. If a bird is sitting on the curtain railing and is reluctant to come down, grabbing it and forcing it down is more likely to result in your being bitten. However, asking the bird to 'step up' and having a piece of fresh vegetation for it to nibble on in your lap when you sit down, is a rewarding experience for the bird.

Behaviour training is not that difficult to do. By consistently rewarding and reinforcing desirable behaviours, you will be able to teach your Cockatiel to eagerly present almost any behaviour that you can imagine.

Above: One of the first lessons to teach your bird is to step up onto your finger. Below: Reward your pet with a treat or a preen.

Bathing and Spraying

Bathing, particularly in hot weather, or showering in the mist of a spray bottle is also a favoured—and enriching—activity. If you are introducing your bird to a misting spray for the first time, it may become frightened if the spray is sudden and too direct. It is best to spray into the air alongside the bird where the edge of the mist just drifts across and touches it. If the bird is going to respond it will begin to extend its wings, fluff up and ruffle its feathers and begin to preen. Some birds will even hang upside down in the mist, spreading their wings and tail feathers to catch droplets of water in their plumage. It is best not to persist if your bird does not like the spray as it is meant to be an activity that gives Cockatiels enjoyment as well as being beneficial to their plumage. In this case, you should provide the bird with a shallow bowl for bathing, or even better, a heavily dampened leafy branch to forage in, providing added enrichment.

Talking

Cockatiels vocalise as a means of communicating with their flock and identifying the locations of other members that may be out of their visual range. In the absence of a flock, Cockatiels will often attempt to mimic the sounds that are familiar to them and sounds that their other 'flock' members such as humans and other pets make. If a tune that their owner whistles to them in an effort to communicate is heard often enough, the Cockatiel will attempt to duplicate it as a response. In return, the bird will expect a response from you. Therefore, if you want your bird to develop a vocabulary, you need to be prepared to spend time talking to your bird and being ready to reply as soon as the bird makes an attempt at communication. Birds will often talk more when you are out of the room or when the lights are turned out suddenly because they want to attract your attention and get you to return to them, or because they feel insecure in the dark and want reassurance.

Although it may seem like simple repetition that could be provided by a CD or a radio, these will not respond in the same way at the appropriate time. I also have some concerns that the repetitive nature of sounds on a CD for long periods of time could be stressful for the birds, although this may be just a personal feeling on my part based on the fact that my sanity would be put to the test if I had to listen to such repetitive sounds all day.

Talking ability will vary amongst individuals, but in general, cocks seem to be more talented at talking than hens. Some Cockatiel owners say that hens do not talk at all but some hens will definitely learn to imitate different sounds.

Household Hazards

Sudden loud noises can alarm your bird, particularly telephones, doorbells and barking dogs. Stoves, sinks and refrigerators are potential death traps for pet Cockatiels. If possible, draw the curtains and close doors to the kitchen and toilet areas while your bird is out of its cage. Potentially poisonous house plants, a pet cat or dog and the fumes from non-stick frying pans and toxic insect sprays are just a few other hazards to watch out for. The most important thing to remember for your pet's safety is simply to be aware of your bird and what it is doing while it is out of its safe cage and playing in your environment.

Wing Clipping
Comment by Dr Bob Doneley

Wing clipping can be defined as the practice of clipping (trimming, cutting) the feathers on a bird's wings so as to limit its ability to fly. The decision to clip your pet Cockatiel's wings is a personal one and must be weighed against the advantages and disadvantages that can be associated with this practice.

You should never rely on a wing clip to prevent your bird from flying away. A wing clip is an aid in controlling and training your bird—but it is not the ultimate flight deterrent! Being small light birds, Cockatiels require only a few flight feathers to grow back to achieve enough flight to get up into a tree where they cannot be reached.

A wing clip should be used as a means of protecting your bird against itself and its environment. It should be performed in such a way that the bird does not injure itself through uncontrolled flight and landing. However, it should not be relied on totally to prevent your bird from escaping!

Should Wing Clipping be Done?

The argument against wing clipping is fairly persuasive. Wing clipping limits a bird's ability to fly, and what could be more natural than seeing a bird fly? This argument has a lot of merit, especially when we are trying to give our birds the best and most interactive environment that we can. Birds do love to fly, it is great exercise for them and it allows them to express themselves and their personalties.

However, the argument for wing clipping is also persuasive. We have taken birds out of a natural environment, and we therefore have a moral obligation to meet their needs for safety and security. By limiting a bird's ability to fly, we can protect it from an environment that it has little control over—a household full of potential toxins and other hazards, or an outside environment full of predators just waiting for an escaped pet bird with little notion of self-defence.

Wing clipping can also be an aid to training a bird as it can increase its concentration span. However, newer training techniques and a better understanding of bird behaviour are quickly making this approach outdated.

Of course, in an ideal situation, a trained and bonded bird in a controlled environment (ie under close supervision with no chance of escape to the outside world) can be fully flighted. However, because this is not possible in most households, clipping a pet Cockatiel's wings may be advisable.

Avian veterinarians report that clipping feathers is not a painful procedure. As long as the feather does not have blood in the shaft (see *Blood Feathers* on page 82), cutting it is no more painful than cutting your hair or fingernails. A feather is a structure made of keratin—cutting it will not hurt.

How Should a Wing Clip be Done?

Every week veterinarians see a few birds that have been injured because of badly done wing clips. Either the owner, the breeder or the pet shop has clipped the wings the wrong way, and the result is broken wings and legs, bruised chests, damaged feathers, and so on. These are unnecessary and avoidable injuries.

A bird's wings should not be clipped until it has learnt to fly (and land) properly and safely. So young chicks should be allowed to fly around the nursery, developing their flight muscles and getting rid of some baby fat. At the same time, they are learning how to control themselves in flight, and how to land without hurting themselves.

Many bird owners, veterinarians and breeders advocate clipping just one wing. The theory is that the bird is unbalanced, flies in a spiral, and therefore cannot fly away. Unfortunately the loss of balance also means that the bird loses the ability to land safely, and many of the injuries described above result from these awkward landings (more like a crash, really). Therefore, clipping both wings in a symmetrical manner is recommended. This leaves the

The stiff primary feathers at the end of the wing provide 'lift and thrust', allowing the bird to take off, go forward, and gain height. Five to seven primary feathers can be cut to reduce the bird's ability to take off and gain height.

bird with some control over its flight and, more importantly, its landing.

If you extend a bird's wing, you will see two layers of feathers—the flight feathers and the coverts. The coverts are the small, soft feathers, lying over the top of the other feathers. The flight feathers are the long feathers along the edge of the wing, and can be divided into primaries and secondaries. The primaries are the longer, stiffer feathers at the end of the wing. They provide the lift that the bird needs to fly. The secondaries are the shorter, softer feathers closer to the body. They act as brakes or flaps, allowing the bird to slow its descent and land safely. So we obviously need to clip the primaries, but leave the secondaries alone.

Blood Feathers

Birds replace their old feathers by moulting them out. This occurs all year round, with only a few feathers being replaced at any one time. New feathers grow from follicles in the skin. They start as a tubular structure encased in a keratin sheath. This structure is full of blood vessels, feeding the growing feather. As the feather matures the blood vessels retract, the sheath comes off and the feather unfolds. However, prior to this event the new feather, known as a blood feather, is very fragile and easily damaged. When this occurs the feather can bleed profusely. While it is unlikely that a healthy bird could bleed to death from such an injury, it is painful and unsightly. Therefore, any wing clip must be performed so that blood feathers are given some protection by mature feathers alongside them.

If you cut a growing feather or if it is damaged accidentally, the feather will bleed heavily. This needs to be stopped as soon as possible. If the bleeding cannot be stopped by applying pressure, the feather will need to be pulled out. An avian veterinarian can do this for you. However, the bird will lose too much blood if you cannot get to a veterinary clinic immediately. If you are going to attempt this procedure yourself, the wing should be supported while the bleeding flight feather is removed with a pair of needle-nosed pliers. It is important to have someone help you in this situation.

Wing Clip Patterns

The requirements for a wing clip are:
- it should be done on both wings;
- only the primary flight feathers should be cut; and
- the clip should provide protection for new blood feathers.

Obviously, the common practice of taking a pair of scissors and cutting all the flight feathers (primary and secondary) at a point level with the skin does not meet these requirements!

A clip pattern I use involves cutting the last 5–7 primaries approximately 50–100mm beyond the edge of the coverts. This takes away the bird's ability to lift, so that while it can fly, it cannot rise. Leaving the secondaries allows for a graceful, controlled landing. By cutting the primaries at this length, there is enough of the shaft of the mature feathers to protect any emerging blood feathers. Therefore this clip meets the requirements for a clip mentioned above. Simply take a pair of scissors, decide where to cut and then make a clean cut along the dotted line.

Who Should Perform This Procedure?

Ideally your bird's wings should be clipped by a veterinarian. However, if you insist on performing the wing clip yourself, then you should adhere to the wing clipping procedures outlined above. If your Cockatiel objects to being restrained while having its wings clipped, you will need a second person to assist you. To avoid future resentment towards you it is advisable to have the other person wrap the bird loosely in a towel and restrain it, while you clip the bird's wings. It is worth remembering, however, that clipping your bird's wings while it is restrained can be a major precipitating factor in the breakdown of the relationship between you and your bird. This is a strong case for the alternative—to have your bird's wings clipped by a veterinarian.

How Long Does a Clip Last?

Because the feathers are being replaced on a continuous basis, all wing clips are invariably short term. The clipped feather will moult out, and be replaced by a new feather that will grow to the normal full length. When enough clipped feathers are replaced, the bird's ability to fly normally is restored. How long this will take depends on the stage of the moulting cycle that the bird is in. So a clip may last for several months, or it may only last a few weeks. Your responsibility as a bird owner is to monitor your bird's wings and flying ability, and have the clip repeated when required.

Restraint Alternatives

Using a bird harness will allow people who do not want to clip their Cockatiel's wings to take the bird outside with them without fear of losing it if it is startled. Unlike leg chains that are potentially harmful for the bird, harnesses are safe and comfortable if properly fitted. Some birds may object to the harness initially and therefore putting it on might result in the handler getting bitten and the bird becoming stressed. If this persists I would not opt for this solution as it may develop into a situation where the bird fears and distrusts its owner. This would be more detrimental than not being able to walk outside with your bird.

Harnesses are a safe restraint option.

Avoiding Behavioural Problems
Comment by Jim McKendry

Like all parrot species, Cockatiels present the potential for learning a range of behaviours that can become problematic in the pet home. Managing behavioural problems once they have become established can be challenging. Prevention is, therefore, the key to developing a successful, long-term relationship with a pet Cockatiel. Avoiding behavioural problems can be achieved by setting yourself up to succeed with careful consideration of the following.

Selecting the Right Cockatiel

The intention of all pet parrot owners is to have a tame and interactive bird. When choosing a Cockatiel as a pet you should spend time interacting with the birds available for sale. Choosing an individual that is already comfortable stepping onto a human hand and allows preening is a great start. Avoid purchasing a bird that displays obvious signs of reluctance to interact with humans and withdraws from contact. This may seem like commonsense, but most of the problems that people have with a pet Cockatiel start with a poor choice of bird as a pet. Only purchase a fully weaned, independent and confident bird that is not begging for food. It should be well socialised,

Tame birds need equal quality time with their owners or jealousy may arise.

Page 83

interacting positively with any other Cockatiels present, and with humans. Cockatiels are the most commonly available handraised pet parrot. If you cannot find a bird that you are comfortable with then wait until one becomes available—it won't take long and will be well worth it!

Food for Thought
Purchasing a Cockatiel from a seller who has taken the time to ensure that the bird is eating a varied diet, including formulated pellets, makes it easier to identify a food treat that you can use in the lifelong training of the bird. As an example, your Cockatiel might be particularly keen on a particular food item as a treat. Because this Cockatiel eats a wide variety of other foods, you can keep this favourite item out of its regular daily diet and instead, offer it at times when you would like to interact with your Cockatiel.

By using a highly valued food item as a reward to reinforce simple desired behaviours, such as stepping up and returning to the cage, everyone in the family can develop a similar, positive association with the Cockatiel. The basic principle here is to establish a consistent routine of positively reinforcing desirable behaviours and then maintain this throughout the life of the bird. Never take basic handling behaviours for granted and always actively reinforce these. This is essential in avoiding problems with handling your Cockatiel as it matures. If you have established a routine of always rewarding your Cockatiel with a highly valued treat each time it responds to a behavioural cue, even for simple cues such as 'step up', then your Cockatiel should maintain its motivation to interact with you and respond positively to you throughout its life. A food reward is universal and helps to build the same positive associations with everyone in the environment. Your pet Cockatiel can therefore learn to value interactions with all household members equally, as the reward for doing so is not as variable as other forms of reinforcement.

Keep in mind that if your Cockatiel can eat as much of its favourite food treat within its cage, then you have reduced its motivation to interact with you to receive that treat. At the other end of the scale, no food item that your Cockatiel is dependent on to maintain a healthy weight should be withheld for training. If all your Cockatiel will eat is seed then obviously you need to work on getting it to eat a varied diet first before you can start isolating food groups to use as rewards. Food rewards for your pet bird should always be in addition to its otherwise complete daily diet. Finding the balance is one of the challenges of building and maintaining positive relationships with your Cockatiel.

The Environment

Behavioural problems associated with boredom are common in pet parrots. The key to avoiding such issues is to provide an environment that is stimulating and enriched. Enrichment opportunities both within and outside of the enclosure area may help to reduce behavioural issues such as excessive vocalisation and feather picking. Simple strategies such as providing a ground-foraging tray that enables the Cockatiel to search for food treats and small seeds, in the same way that it would in the wild, provide great stimulation and a diversion from needing to be with the human carer. All aspects of the environment should be carefully considered, from perching and food bowl placement to the types of natural and artificial enrichment on offer. If you can create a 'habitat' that offers numerous interesting enrichment opportunities for your Cockatiel then you will be on the right track to avoiding behavioural problems.

The most effective component of an enrichment strategy is being creative with the way that you feed your Cockatiel. This is easily achieved by managing the feeding of your Cockatiel so that it is encouraged to explore and interact with its environment. Providing all of the food for the day in a single dish can lead to boredom and a subsequent increase in behavioural problems.

Be creative and do not be afraid to make regular changes to the environment of your pet Cockatiel to keep things interesting. Vary the position of food bowls as well as the types of food on offer. Integrate natural foraging enrichment in the form of safe, non-toxic leafy and flowering branches from common native species such as *Eucalyptus, Corymbia, Banksia, Melaleuca, Callistemon, Acacia* and *Grevillea* spp. Regular opportunities to bathe also provide another enriching experience that helps promote normal self-maintenance behaviours, such as preening.

Lifelong Education

Your Cockatiel will depend on you to provide the learning experiences that it needs to adapt to life in your home. Almost all of the behavioural problems that we encounter with pet Cockatiels are 'learned' behaviours that have been reinforced by the human carer. Ultimately we have the greatest influence on the behavioural outcomes of pet Cockatiels through the way that we interact with them.

Developing deeper understandings about the correct way to interact with your Cockatiel involves engaging in education opportunities for yourself, such as reading articles on parrot behaviour, attending workshops and accessing resources such as DVDs recommended by **ABK Publications**. This education then filters down to your Cockatiel through your being empowered with the knowledge to develop an optimum environment in response to the needs of your pet bird.

Ten Ways to Success

Finally, you should reflect carefully on the following 10 elements for successfully avoiding behavioural problems developing in your pet Cockatiel. These have been adapted from an excellent paper, originally written by avian trainer, Steve Martin, and behaviourist, Susan Friedman PhD (2004).

Build confidence through repetition of positive learning experiences.

- **Focus on 'observable behaviours'.** Develop an awareness of the body language of your Cockatiel. Create links between the way your Cockatiel is behaving, its body language that you can observe and the interactions that are occurring in its environment. If you can link the three together, then you have a great start in identifying why problem behaviours might be occurring and how you can make changes to the environment to reduce those behaviours in the future.

- **Accept responsibility** for both the good and undesirable behaviours that your parrot performs. Because your behaviour always plays a role you might need to modify how you are behaving with your Cockatiel before you can expect any behaviour change from him.

- **Create 'partnerships' with your parrot**. Give your Cockatiel the power to make decisions and set up opportunities for the bird to learn from these experiences.

- **Be sensitive to the need for two-way communication** during the learning process. You need to respond to the behaviour of your parrot just as much as you are seeking to develop responses from him to you. Most circumstances that involve a Cockatiel owner being bitten by their pet occurred because they failed to be sensitive to the communication signals that their bird presented to them prior to the bite. Know when to step back, take a break and set yourself and your Cockatiel up for positive interactions at a later time if his behaviour indicates that he may be aggressive towards you.

- **Ask empowering questions** when you are dealing with behaviours. The two most empowering questions for the parrot owner will be: 'What is the motivation for the behaviour?' and 'How does this behaviour apply to parrots in the wild?' The solution to managing those behaviours effectively lies in the answers to those empowering questions.

- **Set your Cockatiel up to succeed** by carefully arranging the environment to increase the opportunity for learning success. The behaviour of your Cockatiel functions in direct response to environmental stimuli. What can you change in the environment to make undesirable behaviour more difficult and desirable behaviour easier for your Cockatiel?

- **Build confidence** through repetition of positive learning experiences. The more consistently you positively reinforce your Cockatiel's desirable behaviours, the greater the likelihood of your Cockatiel maintaining desirable behaviours and interactions with you.

- **Provide short windows of opportunity** to receive positive rewards. This enhances the response time for desired behaviours and is a key aspect of training. As an example, if your Cockatiel is flighted then it is essential that it is trained to come to you on cue. Providing short windows of opportunity to respond to your cue and receive the consequential reward helps to shorten and strengthen the response time of the desired behaviour.

- **Train your Cockatiel at its pace**. Develop a sensitive awareness of the time of day, situation and location where you will best be set up to succeed. Always aim to end on the delivery of a positive reward and avoid extending training sessions beyond the level of desired responsiveness from your parrot.

- **Develop a 'teamwork' approach**. This is particularly important in family settings where long-term receptivity to multiple humans in such an environment is desired from companion parrots. Work with each other to formulate consistent cues, handling patterns, reward delivery and interactions with your bird.

Using these 10 foundations for successful behaviour management will help both you and your new pet Cockatiel achieve a lifelong friendship.

COLOUR MUTATIONS AND GENETICS

Paleface Grey Pied (left) and Whiteface Grey Pied Cockatiels.

COLOUR BREEDING

Prior to the breeding season, mutation hens and cocks should be housed separately to prevent cocks from mating with genetically unsuitable mates. Hens can lay fertile eggs for up to two weeks after copulation and possibly even longer. In addition, many hens will commence excessive egg laying out of season when housed with cocks and this will cause additional stress to the hens. To obtain accurate and guaranteed results during the breeding season only one pair should be housed in each aviary.

Breeding for specific Cockatiel mutations involves three critical areas of administration apart from appropriate management of your collection. These critical areas of identification, record keeping and a basic understanding of genetics are outlined below.

Cinnamon Suffused Cockatiel hen.

Identification of Breeding Stock and Offspring

Your initial stock should be close banded, but if not, the birds can be fitted with coloured plastic split rings. There are both benefits and disadvantages to closed banding chicks. For breeders of colour mutations it is almost impossible to keep track of the genetic background of birds without some form of identification. Microchipping is replacing banding for larger pet birds and those of considerable value, but as yet it is not practical for small birds and the cost is a consideration for breeders of less expensive species. Another disadvantage is that there is no visual means of identifying birds easily as a scanner is required to read the chip. Some breeders use coloured bands so that they can identify a particular bird easily from a distance without the necessity of catching the bird.

The identification colour or band numbers should be recorded and a visual description of the birds noted. When their young are old enough they should also be banded and any additional genetic information recorded. For instance, if a Lutino chick is unexpectedly hatched from a Normal cock, you can note that the cock is split for Lutino. In future, a more appropriate mate can then be chosen for that particular cock. You will also know that his male offspring will possibly carry forward this gene and any purchaser should be advised of this.

Banding

Chicks may be close banded anywhere from between five days to two weeks of age, depending on how rapidly they are growing. Closed banding simply involves holding the three longest toes together pointing forward. The small, fourth toe is held back out of the way while the ring is slipped over the first three toes. The ring is placed past the ball of the foot until it catches on the back toenail. The back toe is then gently levered forward through the ring by placing a toothpick between the leg and the back toe, in front of the ring. When chicks are banded at the appropriate time the ring will not have to be forced over the foot but there will not be an excessive amount of space either.

Although the chicks' toes are very flexible at this time care must be taken not to harm the chick in any way. Banded chicks should be checked for several days following banding to ensure that the band has not fallen off or slipped up the leg.

A variety of numbered band styles and sizes from 5.3–6.4mm are available. Styles include split, open, closed and wraparound clips made of stainless steel, anodised aluminium or plastic. The main disadvantage of using leg rings is that birds can get caught on objects in the aviary, potentially causing them serious injury, even death. If birds are being bred for the pet market, it may not be necessary to band chicks. However, many lost birds have been reunited with their owners because they have been traced through the bird's leg band. Split bands are far more likely to catch on objects in the aviary and are therefore not recommended.

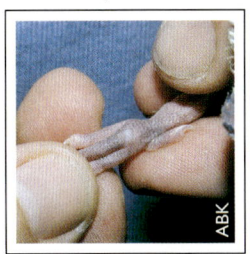

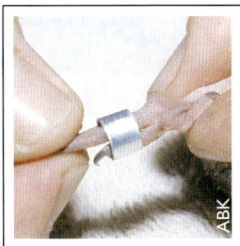

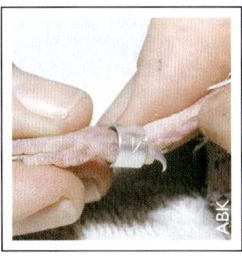

The leg band or ring is placed over the three longest toes, then gently moved down until it meets the fourth toe. Gently pull the fourth toe through the ring until it is located correctly on the chick's leg.

Accurate Record Keeping

All pairings, successful or otherwise, should be recorded in a breeding register. This will include details of the parents, the number of eggs laid, the date the first egg was laid, the date each chick was hatched, as well as the genetic description of the chick and its band number. The register will not only provide the identification of the birds that you have bred but will also assist you to determine the success of individual pairs or specific lines of breeding. The information should indicate if the eggs are overdue in hatching and enable you to make comments on other breeding aspects such as poor parenting skills or hatching difficulties.

Understanding Genetics

When obtaining new stock you may have to rely on the knowledge of the person supplying the birds and the hope that they understand genetics well enough to guarantee the true genetic background of the birds. Frequently, the latter does not apply, resulting in unsuitable pairings and many unexpected breeding results. The confused breeder may then pass incorrect information to the next purchaser based on the inaccurate results obtained.

For this reason, a modest understanding of genetics is required. It is by far the biggest hurdle that most people face when breeding mutation birds, particularly when dealing with the pairing of multi-mutation birds. Breeders may be provided with a list of breeding expectations from a pair that they purchase. However, without an

Whiteface Grey Cockatiel hen.

understanding of the principles of genetics involved, they will often run into difficulties with their next generation of young. Calculating the genetic outcome of a pair of multi-mutation birds, however, is really no more difficult than calculating a simple pairing—if you understand the principles.

It is generally understood that the physical characteristics of all living organisms are passed from parent to offspring by the process of gene inheritance. The basic principles of genetics are all one needs in order to understand Cockatiel mutations.

As with humans and other animals, the individual characteristics of any bird are determined by the formation of the first cell in the reproductive process resulting from the combination of a male cell (sperm) and a female cell (ovum). Each of these cells, specific to reproduction, carries only half the genetic material found in the cells of most living organisms. Their combination results in a cell carrying a full complement of genetic instructions, half from the cock and half from the hen. This cell divides and multiplies as the embryo grows. Within the nucleus of each cell, this genetic information is carried on strands of DNA. These strands, referred to as chromosomes, are linked together in pairs, one from the hen and one from the cock. Situated along the length of the chromosomes are genes in corresponding positions forming pairs.

Genes determine our sex and physical appearance. When a pair of genes are identically matched, such as in the case of a child from blue-eyed parents, the offspring will demonstrate this characteristic. However, if one parent has blue eyes and the other brown eyes, the brown gene is a dominant characteristic and will mask the recessive blue gene, resulting in a child with brown eyes.

The appearance of a new mutation in a species is of interest to both the experienced breeder and the novice bird keeper alike. And as more people become involved in the breeding of Cockatiel mutations, new mutations and combinations will continue to appear.

The decision to breed mutations carries with it the responsibility to preserve the charming wildtype (Normal) Cockatiel. High quality Cockatiel mutations cannot be developed and maintained without also breeding high quality Normal birds. This should be a natural and important part of every specialist breeder's breeding regime.

Paleface Cinnamon Pied Cockatiel cock.

Paleface Grey Cockatiel cock.

Edged Dilute Cockatiel cock.

BASIC GENETICS

Autosomal Recessive
Figure 1
Example of an Autosomal Chromosome Pair

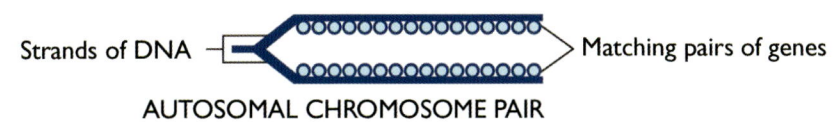

Strands of DNA ——— Matching pairs of genes
AUTOSOMAL CHROMOSOME PAIR

The wildtype Normal Grey Cockatiel carries a large number of 'normal' genes in pairs, that together control all the individual steps for normal pigmentation. Occasionally, the composition of a gene will be altered spontaneously, creating a mutant gene. These mutant (colour) genes are almost always recessive to the wildtype gene and are therefore referred to as autosomal recessive, unless the spontaneous mutation occurred on the chromosomes that determine sex, in which case they are called sex-linked recessive. Recessive genes can be inherited for generations without becoming visible until two birds carrying the same mutant gene mate together. This is generally the result of random inbreeding in uncontrolled aviary situations.

For the purpose of simplifying our explanation of recessive inheritance, birds that have originated from wildtype (Grey) parents and are not carrying any mutant genes will be referred to genetically as 'NN'. This is an abbreviation of the fact that the bird has received only wildtype genes for colour from both parents. This amounts to about 30 pairs of different genes controlling all the various aspects of plumage pigmentation. Geneticists usually list only those that are relevant to a particular discussion. In the following examples we will use 'NN' to symbolise just one generic pair of wildtype genes and 'n' to symbolise a mutant gene of this pair.

Figure 2
Example of an Autosomal Chromosome Pair of Normal Colour = NN

Strands of DNA ——— Matching pairs of genes
AUTOSOMAL CHROMOSOME PAIR

● Gene for the normal colour N
This bird is visually Normal NN

The pairing of two Normal birds is charted as follows:

Figure 3
Normal Paired with Normal = NN x NN

		COCK	
		N	N
HEN	N	NN	NN
	N	NN	NN

All of the offspring are visually and genetically Normal NN.

This pairing indicates that whichever way the chromosomes and the genes that they are carrying link up, the resulting offspring will be Normal—both in appearance and genetically.

Figure 4
Example for a Visually Normal Bird split for Mutation Colour = Nn

Strands of DNA — Unmatched pairs of genes

AUTOSOMAL CHROMOSOME PAIR

● Gene for the normal colour N
○ Gene for the mutation colour n

This bird is visually Normal but split for the mutation colour Nn

Birds carrying a mutant gene are referred to as splits and are abbreviated 'Nn'. The large 'N' refers to the Normal gene and the small 'n' refers to the mutant gene.
The pairing of two split birds is charted as follows:

Figure 5
Split x split = Nn x Nn

COCK

		N	n
HEN	N	NN	Nn
	n	Nn	nn

25% of the offspring are visually and genetically Normal NN.
50% are visually Normal but split for the mutation colour Nn.
25% are visual mutations nn.

This pairing indicates that the genes can link up in four possible combinations. One possibility results in a bird that has not inherited the mutant gene at all and another is a bird that has inherited a mutant gene from both parents and therefore becomes visually different from the Normal bird.

Figure 6
Example of Visual Mutation = nn

Strands of DNA — Matching pairs of genes

AUTOSOMAL CHROMOSOME PAIR

○ Gene for the mutation colour n
This bird is a visual mutation nn

The other two possibilities involve birds that have inherited one mutant gene each. These birds are also splits but are visually Normal because the Normal gene dominates their appearance. It is not possible to tell which birds are carrying the mutant gene and therefore all offspring that are normal in appearance from this pairing are referred to as 'possible splits'. If an 'nn' visually mutated bird is mated to an 'NN' the pairing is charted as follows:

Figure 7
Normal x Visual Mutation = NN x nn

COCK

		N	N
HEN	n	Nn	Nn
	n	Nn	Nn

All of the offspring are visually Normal but split for the mutation colour Nn.

All birds will be visually Normal but all birds must carry a mutant gene and are therefore definite splits.

If an 'nn' visually mutated bird is mated to an 'Nn' split bird, the pairing is charted as follows:

Figure 8
Mutation x split Mutation = nn x Nn

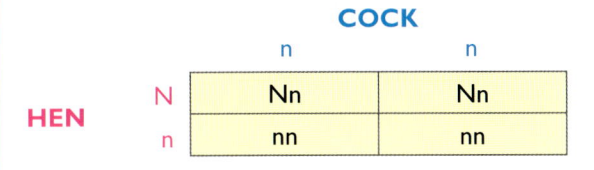

COCK

		n	n
HEN	N	Nn	Nn
	n	nn	nn

50% of the offspring are visually Normal but split for the mutation colour Nn. 50% are visual mutations nn.

In this case all birds have either inherited one mutant gene or two and are either visual mutations or definite splits.

If an 'nn' visually mutated bird is mated to another 'nn' mutated bird, the pairing is charted as follows:

Figure 9
Visual Mutation x Visual Mutation = nn x nn

COCK

		n	n
HEN	n	nn	nn
	n	nn	nn

All offspring are visual mutations nn.

In this case all birds are visual mutations.

With autosomal mutations both hens and cocks can be split for a mutation colour. As indicated in the above charts you can change the sex of the split and the coloured bird, ie the combination of the split cock and a visual mutation hen can be changed to a split hen and a visual mutation cock, but the resulting offspring remain genetically the same. In Cockatiels, the mutations of Ashen Fallow (European 'Recessive Silver'), Bronze Fallow, Dilute ('Pastel' Silver), Edged Dilute ('Silver Spangle'), USA 'Emerald', Faded ('West Coast Silver'), Paleface ('Pastelface'), Pied, Suffused (Australian 'Olive') and Whiteface are autosomal recessive mutations. Both hens and cocks can be split for these mutations. When charting the breeding of these birds, many people assign unrelated abbreviations to these mutations, which causes the process to be more difficult to understand. As our understanding of mutations increases, we are able to better identify the genes responsible for different colours. There are now international attempts to standardise names and symbols used. This is discussed in more detail in later chapters of this section. When dealing with different pairs of genes, each pair of genes is given a different symbol. For each pair recessive genes are symbolised with lower-case letters and therefore the corresponding wildtype Normal gene uses the upper-case letter. European breeders prefer to use '+' as superscript to symbolise the wildtype gene. For simplification in the following tables we will only use the following abbreviations:

P = wildtype gene   p = Pied
F = wildtype gene   f = Fallow
S = wildtype gene   s = Silver

Sample Pairings

Figure 10
Normal split Pied x Normal split Pied = Pp x Pp

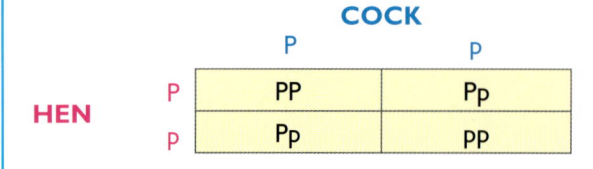

		COCK	
		P	p
HEN	P	PP	Pp
	p	Pp	pp

25% of the offspring are visually and genetically Normal PP.
50% are visually Normal split Pied Pp.
25% are visual Pieds pp.

Figure 11
Normal x Pied = PP x pp

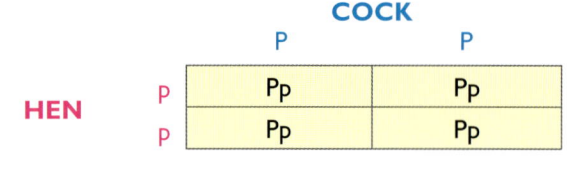

		COCK	
		p	p
HEN	P	Pp	Pp
	P	Pp	Pp

All offspring are visually Normal split Pied Pp.

Figure 12
Pied x Normal split Pied = pp x Pp

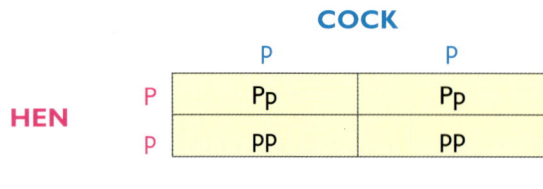

		COCK	
		P	p
HEN	p	Pp	pp
	p	Pp	pp

50% of the offspring are visual Pieds pp.
50% are visually Normal split Pied Pp.

Figure 13
Pied x Pied = pp x pp

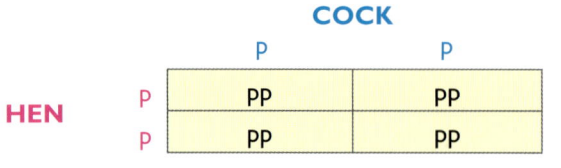

		COCK	
		p	p
HEN	p	pp	pp
	p	pp	pp

All of the offspring are visual Pieds pp.

Calculating Multi-Autosomal Mutations

This is done in exactly the same manner as shown in the above charts. Take for example the pairing of a Silver split Pied cock with a Pied split Silver hen.

Figure 14
Firstly chart the split Pied with the Pied = Pp x pp

		COCK	
		P	p
HEN	p	Pp	pp
	p	Pp	pp

50% of the offspring are visual Pieds pp.
50% are visually Normal split Pied Pp.

Figure 15
Then the Silver with the split Silver = ss x Ss

		COCK	
		s	s
HEN	S	Ss	Ss
	s	ss	ss

50% of the offspring are visual Silvers ss.
50% are visually Normal split Silver Ss.

If you combine these two charts they will read as follows:

Figure 16
Silver split Pied x Pied split Silver = ssPp x Sspp

		COCK	
		sP	sp
HEN	sp	ss Pp	ss pp
	Sp	Ss Pp	Ss pp

25% of the offspring are Silver split Pied ss Pp.
25% are Normal split Pied and Silver Ss Pp.
25% are Pied split Silver Ss pp.
25% are Silver Pied ss pp.

You then simply read off the results—wherever a bird carries two matching symbols it is a visual mutation and where it carries only one it is a split.

Autosomal Co-dominant

Mutant genes can sometimes behave dominantly and suppress the action of the wildtype Normal gene. In birds these genes are usually co-dominant rather than pure dominant. This means that they only partially suppress the wildtype gene and therefore produce two different appearances, the single factor form (one mutant gene and one wildtype gene) and the double factor form (two mutant genes). Co-dominant mutations can be either autosomal co-dominant or sex-linked co-dominant, however the latter are currently unknown in Cockatiels. The European Dominant Edged ('Dominant Silver') and the American dominant Yellowface are two examples of this form of inheritance.

Calculating breeding results is identical to that used for recessive traits, except now the mutation is the uppercase symbol and the wildtype Normal gene is the lower case symbol, or more appropriately for co-dominant mutations an uppercase symbol with a superscript '+' to distinguish it.

Single factor Dominant Edged Cockatiel hen.

Single factor Dominant Edged Cockatiel cock.

Page 95

Sex Chromosomes in Birds

The sex chromosomes in birds are correctly referred to as Z and W, the male being ZZ and the female ZW. Aviculturists have traditionally used the XY designations that apply to mammals which is incorrect because in mammals the XY genotype is male while in birds the ZW genotype is female. There is absolutely *no* relationship between the XY of mammals and the ZW of birds, except that their combination plays a role in determining sex in their respective group of animals. Most of the genes present on the X chromosome of mammals are not present on the Z chromosome of birds and the converse is also true. In fact, the avian Z chromosome carries many of the genes normally present on chromosome 9 of mammalian species, as well as a few genes adopted from a number of other mammalian chromosomes.

It is argued by some breeders that changes like this are too confusing and as such they persist in using the incorrect terminology. Yet, if they are to understand avian genetics, they must also understand the differences between animals and once they are aware of this difference, it cannot be logically ignored. Breeders who adopt the correct nomenclature are demonstrating a better grasp of genetic principles and after all it is only a simple change to understand.

A casual, colloquial use of terms gives only a superficial understanding of genetics that will be full of inaccuracies and confusion. In contrast, with a deeper knowledge of terminology, the more precisely we can apply it and then we begin to fathom the more complex aspects of avian pigmentation and its genetic inheritance. We even learn new facts about the 'simple things' that we previously took for granted.

Sex-linked Recessive

One pair of chromosomes within a cell's nucleus differs from the others. This pair is responsible for determining sex. Cocks have a matched pair similar to their other chromosomes and, like those, they carry a full complement of genes. This matching pair of male chromosomes is represented by the symbol 'ZZ'. Hens do not have a matched pair of sex chromosomes. One of hers is identical to the male chromosome 'Z', but the other is short and only able to carry a few genes concerned with reproduction. This short chromosome is represented by the symbol 'W' and therefore, the female sex chromosomes are represented by the symbols 'ZW'. This differs from mammals, where the female carries a matched set of sex chromosomes and the male carries an unmatched set and the symbols XX and XY are used.

Cinnamon Cockatiel cock.

Figure 17
Sex-linked DNA

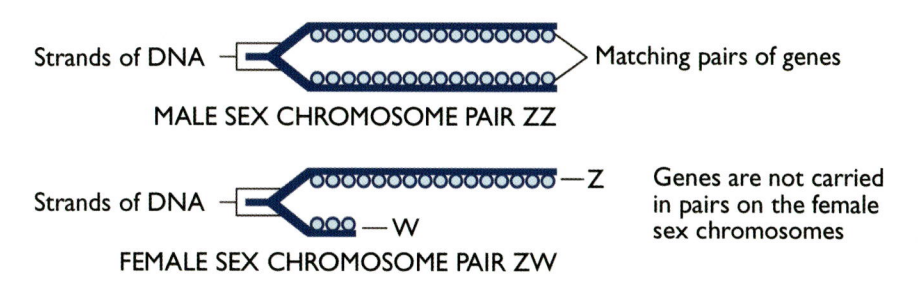

Strands of DNA → Matching pairs of genes

MALE SEX CHROMOSOME PAIR ZZ

Strands of DNA — Z / W

Genes are not carried in pairs on the female sex chromosomes

FEMALE SEX CHROMOSOME PAIR ZW

As mentioned earlier, when a new cell is formed by the union of sperm and ovum, each pair of new chromosomes originates from the combination of one donated by the cock and one by the hen. In all cases the cock donates a 'Z' chromosome but the hen can donate either a 'Z' or a 'W'. If the newly formed cell receives a 'Z' from the hen, it must develop into a cock 'ZZ', but if it receives a 'W' it will develop into a hen 'ZW'.

Therefore, it is the hen that determines the ultimate sex of her offspring. The union of male and female sex cells is charted as follows:

Figure 18
Sex Cells Only

		COCK	
		Z	Z
HEN	Z	ZZ	ZZ
	W	ZW	ZW

50% of the offspring are male ZZ.
50% are female ZW.

Figure 18 suggests that two out of every four chicks must be female, however, in reality we know that this is not always the case but merely a statistical average. There is growing scientific evidence that hens can, by an undetermined method, influence the sex of eggs that they produce. This allows them to alter the sex ratios of their chicks under different environmental and other conditions. For instance a hen could produce a higher percentage of ova that predominantly carry the 'Z' chromosome, resulting in a higher percentage of cocks in each clutch.

In some cases, certain mutant genes will appear on the 'Z' sex chromosome. They do not appear on another chromosome and are therefore referred to as 'sex-linked' and are recessive to the dominant normal colour. Because cocks have a matched pair of 'Z' chromosomes, they are able to carry the gene for normal colour opposite a mutant gene, therefore masking the effect. In this case the cock is visually Normal but split for the sex-linked colour. The cock must carry a matched pair of mutation sex-linked genes before he will be the visual mutation colour.

In contrast, if the hen carries the same sex-linked mutant gene on her 'Z' chromosome she will be the visual mutation colour with only one gene present. The short 'W' chromosome is unable to carry the gene for normal colour opposite and therefore there is nothing to mask the effects of the mutant gene. For the same reason hens cannot be split for a sex-linked mutation colour. If the mutant gene is present on the 'Z' chromosome, then it must show as a visible change in the hen's colour.

Figure 19
Sex-linked Mutation DNA

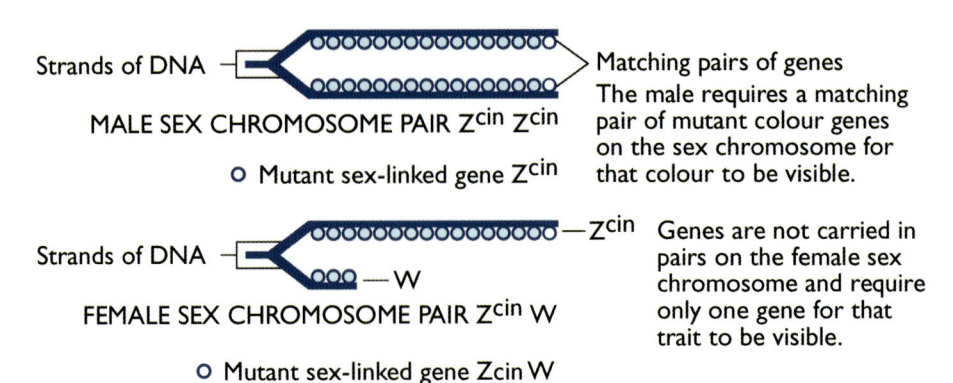

MALE SEX CHROMOSOME PAIR $Z^{cin}\ Z^{cin}$

Strands of DNA

Matching pairs of genes
The male requires a matching pair of mutant colour genes on the sex chromosome for that colour to be visible.

○ Mutant sex-linked gene Z^{cin}

Strands of DNA

FEMALE SEX CHROMOSOME PAIR $Z^{cin}\ W$

— Z^{cin}
— W

Genes are not carried in pairs on the female sex chromosome and require only one gene for that trait to be visible.

○ Mutant sex-linked gene $Z^{cin}\ W$

In Cockatiels the mutations of Lutino, Cinnamon, Pearl, Platinum, Pewter and Yellowcheek are sex-linked. When charting the possible offspring from sex-linked mutations use the following abbreviations:

cin = Cinnamon
ino = Lutino
op = Pearl

Because a hen only needs to receive one gene to be visible for a sex-linked colour, it is important to include the symbols for the sex chromosomes when charting the pairings of sex-linked birds. If not, it would be impossible to determine whether a bird displaying only one mutation symbol was a split cock or a visual hen.

A visual Lutino cock is therefore assigned the symbols $Z^{ino}\ Z^{ino}$.
A split Lutino cock is $Z^{ino}\ Z$.
A visual Lutino hen is $Z^{ino}\ W$.

Sex-linked mutations are represented in superscript, above the Z symbol. This immediately indicates that it is sex-linked and differentiates it from an autosomal recessive mutation. In charting sex-linked and multi-mutation pairings, it is not really necessary to use the symbols to represent the wildtype gene as any cocks or hens that do not have a matching pair of mutation symbols or hens that have no individual sex-linked symbols are always visually Normal.

Sample Pairings

Figure 20
Split Lutino cock x Normal hen = $Z\ Z^{ino}$ x ZW

HEN \ COCK	Z^{ino}	Z
Z	$Z\ Z^{ino}$	ZZ
W	$Z^{ino}\ W$	ZW

25% of the offspring are split Lutino cocks $Z\ Z^{ino}$.
25% are Normal cocks ZZ.
25% are Lutino hens $Z^{ino}\ W$.
25% are Normal hens ZW.

Figure 21
Lutino cock x Normal hen = $Z^{ino} Z^{ino} \times ZW$

	COCK	
	z^{ino}	z^{ino}
HEN Z	$Z Z^{ino}$	$Z Z^{ino}$
HEN W	$Z^{ino} W$	$Z^{ino} W$

50% of the offspring are split Lutino cocks $Z Z^{ino}$.
50% are Lutino hens $Z^{ino} W$.

Figure 22
Split Lutino cock x Lutino hen = $Z Z^{ino} \times Z^{ino} W$

	COCK	
	z^{ino}	Z
HEN Z^{ino}	$Z^{ino} Z^{ino}$	$Z^{ino} Z$
HEN W	$Z^{ino} W$	ZW

25% of the offspring are Lutino cocks $Z^{ino} Z^{ino}$.
25% are Normal split Lutino cocks $Z^{ino} Z$.
25% are Lutino hens $Z^{ino} W$.
25% are Normal hens ZW.

Figure 23
Lutino cock x Lutino hen = $Z^{ino} Z^{ino} \times Z^{ino} W$

	COCK	
	z^{ino}	z^{ino}
HEN Z^{ino}	$Z^{ino} Z^{ino}$	$Z^{ino} Z^{ino}$
HEN W	$Z^{ino} W$	$Z^{ino} W$

50% of the offspring are Lutino cocks $Z^{ino} Z^{ino}$.
50% are Lutino hens $Z^{ino} W$.

Calculating Sex-linked Mutations

Cocks can be split for several sex-linked mutations or be visual for one mutation and split for up to four others, visual for two or three mutations and split for one, and so on. Hens can be visual for more than one sex-linked mutation such as Cinnamon and Pearl but cannot be split for sex-linked colours. In the case of a cock that is split for two sex-linked colours, the mutant genes can be carried on the same 'Z' chromosome or he can carry one on each 'Z' chromosome. When charting his potential offspring, two charts must be done in order to cover both possibilities.

Figure 24
Example of Normal split Lutino split Pearl cock x Normal hen
= $Z^{ino} Z^{op} \times ZW$ or $Z^{ino,op} Z \times ZW$

	COCK	
	z^{ino}	z^{op}
HEN Z	$Z Z^{ino}$	$Z Z^{op}$
HEN W	$Z^{ino} W$	$Z^{op} W$

Offspring are:
Normal split Lutino cocks $Z Z^{ino}$.
Normal split Pearl cocks $Z Z^{op}$.
Lutino hens $Z^{ino} W$.
Pearl hens $Z^{op} W$.

	COCK	
	$z^{ino,op}$	Z
HEN Z	$Z Z^{ino,op}$	ZZ
HEN W	$Z^{ino,op} W$	ZW

Offspring are:
Normal split Lutino and Pearl cocks $Z Z^{ino,op}$.
Normal cocks ZZ.
Lutino Pearl hens $Z^{ino,op} W$.
Normal hens ZW.

As Figure 24 indicates, there are eight different genetic combinations possible from this pairing, however all the cocks are visually Normal and it is impossible to tell which are split to the sex-linked colours. Therefore, all the cocks are referred to as possible splits but the hens are either visual mutations or genetically and visually Normal. If the cock was also split for Cinnamon, four charts would need to be done to chart the cock's splits in the following combinations:

Z^{cin} $Z^{ino,op}$
$Z^{cin,ino}$ Z^{op}
$Z^{cin,op}$ Z^{ino}
$Z^{cin,ino,op}$ Z

Ideally, in the pairing of any two birds, the best possible combination is that of a visual mutation hen and a cock that is, at least, split for the same mutation. In this way all the offspring are visual mutations or definite splits. The same applies to multi-mutations. Following then, is a set of charts calculating the offspring of a multi-split cock and a multi-visual hen:

Figure 25
Example of Normal split Lutino, Cinnamon and Pearl cock x Lutino Cinnamon Pearl hen = N/ino/cin/op x cin ino op

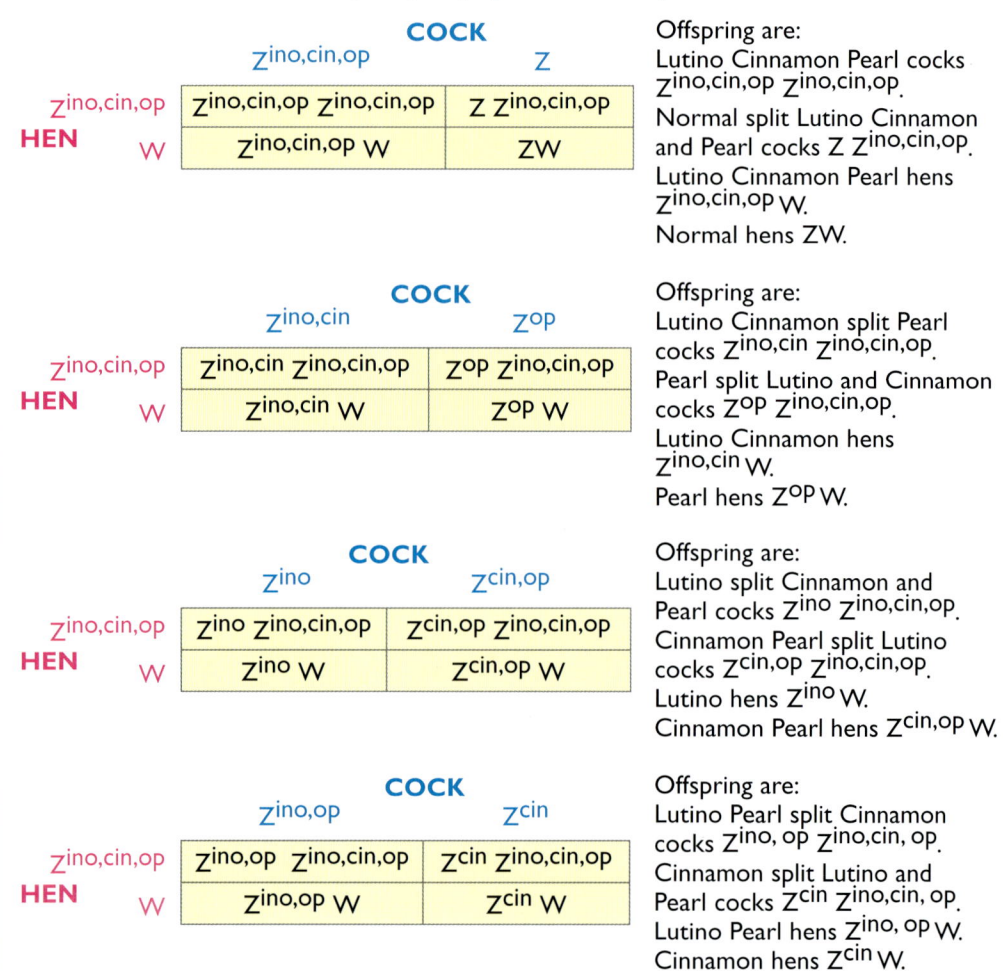

COCK

	$Z^{ino,cin,op}$	Z
HEN $Z^{ino,cin,op}$	$Z^{ino,cin,op}$ $Z^{ino,cin,op}$	Z $Z^{ino,cin,op}$
W	$Z^{ino,cin,op}$ W	ZW

Offspring are:
Lutino Cinnamon Pearl cocks $Z^{ino,cin,op}$ $Z^{ino,cin,op}$.
Normal split Lutino Cinnamon and Pearl cocks Z $Z^{ino,cin,op}$.
Lutino Cinnamon Pearl hens $Z^{ino,cin,op}$ W.
Normal hens ZW.

COCK

	$Z^{ino,cin}$	Z^{op}
HEN $Z^{ino,cin,op}$	$Z^{ino,cin}$ $Z^{ino,cin,op}$	Z^{op} $Z^{ino,cin,op}$
W	$Z^{ino,cin}$ W	Z^{op} W

Offspring are:
Lutino Cinnamon split Pearl cocks $Z^{ino,cin}$ $Z^{ino,cin,op}$.
Pearl split Lutino and Cinnamon cocks Z^{op} $Z^{ino,cin,op}$.
Lutino Cinnamon hens $Z^{ino,cin}$ W.
Pearl hens Z^{op} W.

COCK

	Z^{ino}	$Z^{cin,op}$
HEN $Z^{ino,cin,op}$	Z^{ino} $Z^{ino,cin,op}$	$Z^{cin,op}$ $Z^{ino,cin,op}$
W	Z^{ino} W	$Z^{cin,op}$ W

Offspring are:
Lutino split Cinnamon and Pearl cocks Z^{ino} $Z^{ino,cin,op}$.
Cinnamon Pearl split Lutino cocks $Z^{cin,op}$ $Z^{ino,cin,op}$.
Lutino hens Z^{ino} W.
Cinnamon Pearl hens $Z^{cin,op}$ W.

COCK

	$Z^{ino,op}$	Z^{cin}
HEN $Z^{ino,cin,op}$	$Z^{ino,op}$ $Z^{ino,cin,op}$	Z^{cin} $Z^{ino,cin,op}$
W	$Z^{ino,op}$ W	Z^{cin} W

Offspring are:
Lutino Pearl split Cinnamon cocks $Z^{ino,op}$ $Z^{ino,cin,op}$.
Cinnamon split Lutino and Pearl cocks Z^{cin} $Z^{ino,cin,op}$.
Lutino Pearl hens $Z^{ino,op}$ W.
Cinnamon hens Z^{cin} W.

Calculating Combinations between Sex-linked and Autosomal Mutations

Now we get to the tricky part! However, it is not really as difficult as it might seem. In addition to being visual sex-linked colours or split for that colour in the case of cocks, a bird can also carry an autosomal mutation on another set of chromosomes. This will be visual in addition to the sex-linked colours if the bird has a matched pair of chromosomes or split if it only carries one mutant gene. An obvious example is Pearl Pied where the visual effects of both the autosomal recessive mutation Pied and the sex-linked mutation Pearl are visible at the same time. When calculating the breeding results of combinations between the two types of mutations, the principle is exactly the same.

All that is required is to chart the sex-linked and autosomal mutations on the same chart, doing as many charts as is required to cover the different possibilities.

Figure 26
Example of Normal split Pearl split Pied cock x Pearl hen
$$= Z^{op} Z, Pp \times Z^{op} W, PP$$

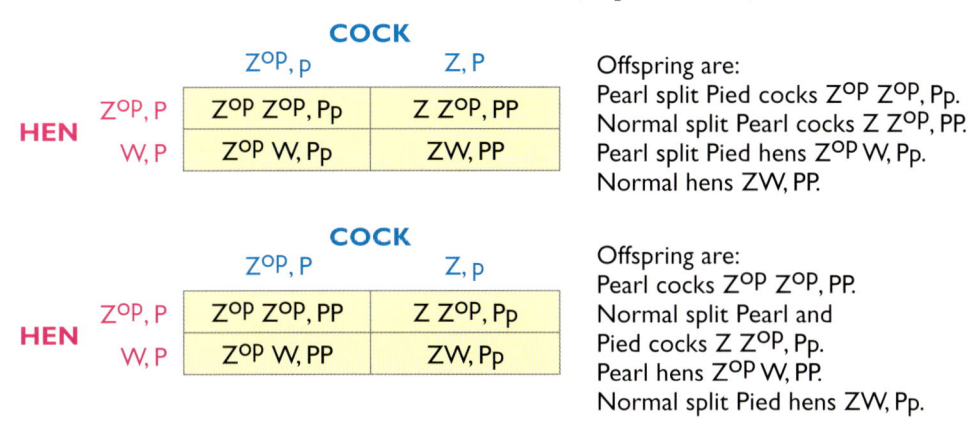

COCK

HEN		Z^{op}, p	Z, P
	Z^{op}, P	$Z^{op} Z^{op}$, Pp	$Z Z^{op}$, PP
	W, P	Z^{op} W, Pp	ZW, PP

Offspring are:
Pearl split Pied cocks $Z^{op} Z^{op}$, Pp.
Normal split Pearl cocks $Z Z^{op}$, PP.
Pearl split Pied hens Z^{op} W, Pp.
Normal hens ZW, PP.

COCK

HEN		Z^{op}, P	Z, p
	Z^{op}, P	$Z^{op} Z^{op}$, PP	$Z Z^{op}$, Pp
	W, P	Z^{op} W, PP	ZW, Pp

Offspring are:
Pearl cocks $Z^{op} Z^{op}$, PP.
Normal split Pearl and
Pied cocks $Z Z^{op}$, Pp.
Pearl hens Z^{op} W, PP.
Normal split Pied hens ZW, Pp.

In this case a single autosomal mutation gene is introduced to a cock carrying a single sex-linked gene. These can be inherited either together or separately and therefore two charts need to be done to cover these possibilities.

Figure 27
Normal split Pearl split Pied cock x Pearl split Pied hen
$$= Z^{op} Z, Pp \times Z^{op} W, Pp$$

COCK

HEN		Z^{op}, p	Z, P
	Z^{op}, p	$Z^{op} Z^{op}$, pp	$Z Z^{op}$, Pp
	W, P	Z^{op} W, Pp	ZW, PP

Offspring are:
Pearl Pied cocks $Z^{op} Z^{op}$, pp.
Normal split Pearl and
Pied cocks $Z Z^{op}$, Pp.
Pearl split Pied hens Z^{op} W, Pp.
Normal hens ZW, PP.

COCK

HEN		Z^{op}, P	Z, p
	Z^{op}, p	$Z^{op} Z^{op}$, Pp	$Z Z^{op}$, pp
	W, P	Z^{op} W, PP	ZW, Pp

Offspring are:
Pearl split Pied cocks $Z^{op} Z^{op}$, Pp.
Pied split Pearl cocks $Z Z^{op}$, pp.
Pearl hens Z^{op} W, PP.
Normal split Pied hens ZW, Pp.

In this case the same single autosomal gene is introduced to the hen. Again, these can be inherited together or can be inherited separately. Therefore, four charts are now required to cover all the possibilities. Firstly chart the different combinations that can occur in the cock while the hen combination stays constant.

Then chart these same combinations with any other positions in which the hen's autosomal recessive genes can be arranged.

Figure 28

	COCK	
	Z^{OP}, P	Z, P
HEN Z^{OP}, P	$Z^{OP} Z^{OP}, Pp$	$Z Z^{OP}, PP$
W, p	$Z^{OP} W, pp$	ZW, Pp

Offspring are:
Pearl split Pied cocks $Z^{OP} Z^{OP}, Pp$.
Normal split Pearl cocks $Z Z^{OP}, PP$.
Pearl Pied hens $Z^{OP} W, pp$.
Normal split Pied hens ZW, Pp.

	COCK	
	Z^{OP}, P	Z, p
HEN Z^{OP}, P	$Z^{OP} Z^{OP}, PP$	$Z Z^{OP}, Pp$
W, p	$Z^{OP} W, Pp$	ZW, pp

Offspring are:
Pearl cocks $Z^{OP} Z^{OP}, PP$.
Normal split Pearl and
Pied cocks $Z Z^{OP}, Pp$.
Pearl split Pied hens $Z^{OP} W, Pp$.
Pied hens ZW, pp.

In this case the Pied has been separated from the hen's visual sex-linked gene and is now inherited by her female offspring. In the case of hens, sex-linked genes cannot alter position because they must always ride on the 'Z' gene. In this particular pairing both hen and cock are only split for the autosomal recessive gene 'Pied'. As you can see by the charts, this results in Pearl and visually Normal birds, some that are split Pied and some that are not, and also some visual Pieds. As it is usually impossible to distinguish which offspring have inherited the gene, all non-visual Pieds must be referred to as 'possible split Pieds'.

Much more complicated pairings can be treated in the same way to supply you with accurate breeding results. Through experience, you will be able to calculate the results of many pairings without resorting to pen and paper. For complex pairings always revert to this method in an effort to avoid missing any possibilities.

If all else fails, try using a computer program to calculate the breeding results. A large range of computer programs designed to do the mathematics for you are now available. All you have to do is enter the correct genetic make-up for the parents and all possible offspring are automatically listed. Some programs also allow you to store all your records electronically and these will even track the genetic make-up of the parents.

Grey Pearl Pied Cockatiel hen.

MUTATIONS

Wildtype Pigmentation

The Cockatiel is one of the few popular parrot species that is not able to produce true optical green and blue colours. This is because the ancestral Cockatiel lost the genes required to produce these structural colours and the ability can never be reclaimed. Genetically, this has made the Cockatiel equivalent to a 'Greygreen' in other species, although the yellow pigmentation has also been largely reduced in distribution within the plumage to leave the wildtype bird pure grey to a large extent, particularly in the cock.

Psittacofulvin is present as yellow pigment in the face mask of the cock and as orange pigment in the cheekpatches. These pigments occupy the same areas but are partly hidden by increased melanin suffusion spread through the face of the hen, whereas the face area of the adult cock is devoid of melanin. Also importantly, there is spread of yellow psittacin pigment through much larger areas of the plumage in the hen, which becomes visible due to an intricate pattern of melanin pigmentation loss from the underside of the wing, tail feathers and lower abdomen.

Normal Grey Cockatiel cock.

Normal Grey Cockatiel hen.

The differences in pigment distribution produce a significant plumage sexual dimorphism in the adult wildtype, which is also visible in different ways in most of the colour morphs available today.

The basic pigment components that the Cockatiel carries in its plumage determine which colour morphs it can produce. Ample melanin production in this species results in very clearly defined melanin-altering mutations from all subcategories (albinism, leucism and dilution). In fact, these are actually easier to differentiate than in a typical 'green' species of parrot where heavy yellow pigment causes many of these colours to look very similar. As such, the Cockatiel probably has the greatest number of established melanin-altering mutations in psittacine aviculture. These include Lutino, Cinnamon, Fallow, Platinum, Pied and all the different 'Silver' mutations around the world, currently numbering another eight.

Psittacin pigment-altering mutations are also possible, due to the presence

Page 103

of yellow and orange pigments in the Normal bird. Surprisingly, despite their reduced distribution in the Cockatiel, this species has four different mutations in this category including Blue (Whiteface), Aqua (Paleface, aka 'Pastelface'), Tangerine (dominant Yellowface) and the sex-linked Yellowcheek genes. Due to the lack of structural colour, the Cockatiel is not able to produce mutations of those colours and therefore cannot produce Dark factor or Violet factor colour morphs. It does, however, have the well-established Opaline mutation, known as Pearl, to round off the range of colours.

Mutation versus Colour

I wish to make it very clear before proceeding that there is a difference between the terms 'mutation' and 'colour'. The word 'mutation' refers to the genetic change that has occurred, not the colour that is produced by that change. It relates directly to the genes involved and their identity, in other words the genotype. 'Colour' is the correct term relating to the phenotype or external appearance of the bird. As such, when I state that the Whiteface Cockatiel is the Blue mutation, I am referring to the gene and its action. I am not maintaining that the colour name must be changed to 'Blue' which clearly would be difficult to apply to a Cockatiel. Yet, if the breeder understands that this Whiteface colour morph is produced by the same gene that produces Blue colour morphs in most other parrot species, they can gain added insight into the nature of the bird they are breeding—and gain the ability to extrapolate technical knowledge across species with the same genetic mutations.

Normal Grey (wildtype) Cockatiel hen (left) and cock.

Whiteface Grey Cockatiel cock. Whiteface is the correct name for the colour, Blue is the correct name for the mutation.

Colour names can be the mutation name and this is common practice in applied genetics. Sometimes, due to species specifics of the phenotype, special names for colours have been adopted within aviculture and this is reasonable. It is only inappropriate when the names used for colours refer to the wrong mutation. In these instances I will state that a colour is incorrectly known by the old name. An example is the 'Recessive Silver' in Europe and the USA which is actually a type of Fallow mutation and the term 'Silver' should be reserved for an entirely different type of mutation known as Dilute. Insistence by breeders in using the incorrect name simply leads to ongoing confusion and difficulty in understanding the nature of the mutations involved. Indeed, continual use of a name known to be wrong implies that the breeder does not understand the meaning of the names they are using.

ESTABLISHED MUTATIONS

Taking into account the genetic potential of the Cockatiel, a large number of the possible mutations have occurred and been established, including a couple that have not so far occurred in other parrot species. The list of established Cockatiel mutations can be divided into the following categories:

Melanin-altering Mutations
Albinistic Genes
- cinnamon
- sex-linked ino
- sex-linked parino (platinum)
- faded
- non sex-linked (nsl) ino
- bronze fallow
- ashen fallow
- pewter (unique to Cockatiels)
- US 'Emerald' (not fully classified at present)

Dilution Genes
- dominant edged
- dilute
- suffused
- edged dilute

Leucistic Genes
- recessive pied

Psittacin-altering Mutations
- blue
- parblue (aqua)
- tangerine (dominant yellowface)
- sex-linked yellowcheek (unique to Cockatiels)

Pattern Mutations
- opaline (pearl)

Potential remains for further mutations in the albinistic subgroup. Currently there is no recognised Pastel mutation (NSL Parino) although there are potential candidates amongst the existing mutations that cannot be test mated until the NSL Ino becomes more widely established. Further distinct Fallow mutations are also possible, in particular a Dun Fallow mutation could be expected to appear in the future.

Dilute mutations are fully established, while leucistic mutations are poorly represented and present the greatest potential for new mutations. To date, there are no established Dominant Pied genes, no Mottled and no

The Pewter mutation has only recently been discovered in Cockatiels and has no equivalent in other parrot species.

Pearl Cockatiel hen. The Pearl colour morph in Cockatiels is produced by the Opaline mutation.

Black-eyed Clear gene (a single gene for a clear phenotype, in contrast to the selected 'clear' Pieds that currently exist).

The psittacin pigment-altering category has potential for further Parblue genes. However, with the restricted presence of these pigments in this species, it may not be possible to identify differing phenotypes for distinct Parblue genes. The sex-linked Yellowcheek mutation is, so far, unique to this species of parrot.

Opaline is the primary pattern mutation currently recognised across parrot species. Others may one day appear but so far no others have come to light in aviculture. Melanistic mutations are also potentially possible, although those currently known in parrots generally 'fill in' areas of plumage devoid of melanin, rather than create a true black specimen. In the Cockatiel the white wing bar could be pigmented, as well as the face in the adult cock.

As mentioned already, the Cockatiel does not retain any potential for structural colour mutations and therefore cannot develop Dark factor, Violet factor, Slate, Khaki or Misty mutations. And of course, the Grey factor gene is already applicable to the entire species and all its colour morphs.

The sex-linked Yellowcheek mutation is, so far, unique to this species of parrot.

UNIVERSAL (WORLDWIDE) MUTATIONS

These mutations and colours have been established for a significant period of time and have been spread to all continents of the globe.

Sex-linked Lutino (Z^{ino})

The common Lutino mutation was first bred in the USA in 1958 by Mr Cliff Barringer (Smith 1978; Sindel & Lynn 1989), established by Mrs E L Moon and is now familiar worldwide. When it first became available to the public in Australia (late 1970s) it was commonly sold as an 'Albino' or 'White', but after a few years breeders realised the inaccuracy of those names and now correctly refer to the colour as Lutino. There is also a second Lutino mutation, the autosomal recessive (NSL) Lutino, which has been bred in Europe but is still rare.

The sex-linked recessive Lutino is produced by a common mutation in parrots of the sex-linked ino locus. This gene plays an important role in producing melanin pigment and when mutated it results in almost complete loss of melanin from the entire body, both the plumage and soft tissue structures including eyes, beak, legs and nails. The eyes appear red from when the chick hatches, although in

Lutino Cockatiel cock.

The Lutino Cockatiel cock has only restricted areas of yellow pigment.

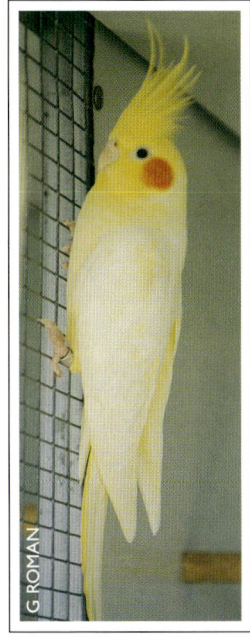

The Lutino hen shows the true spread of yellow pigments in the plumage.

adult birds a slight darkening of the eye colour is apparent, indicating a tiny degree of melanin production. The natal down of the Cockatiel chick is unaffected by the Lutino mutation as it already lacks melanin and appears yellow from hatch.

Removal of melanin from the Cockatiel's plumage reveals the full extent of psittacin pigments throughout the plumage. In the cock, these pigments are largely restricted to the yellow face and bright orange cheekpatches and therefore the rest of the plumage appears white. However, as well as having a yellow face and orange cheekpatches, the hen has a visible, fine pattern of yellow pigmentation under the tail and wings. There is also a faint suffusion of yellow pigmentation through much of the body colour. With the melanin removed, the yellow face and orange cheekpatches of the hen are virtually as bright as those of the cock. The fact that the majority of the plumage appears white, is what caused some of the early confusion amongst breeders about the true identity of this mutation.

Cross and Andersen (1994), in the first edition of this book, add to this description. They note that the long tail feathers and primary flights are usually a lighter shade of yellow than the body. The white wing bar is always retained, although it is not obvious and can be streaked with yellow in brighter specimens. The loss of grey unmasks the bright orange cheekpatch and yellow crest in both sexes. Variation in the size and colour of cheekpatches on individual birds is normal and is not linked to the sex of the bird.

Lutino Cockatiels are notorious for a number of problems. The most obvious is a tendency for developing a bald patch on top of the head behind the crest. This appears to be a genetic trait although it is not caused by the Lutino gene directly. It has merely been bred into Lutino strains through careless selection during the early years of its establishment in this country. It is now a trait which plagues many strains of Lutino but which can be selected against and eliminated with concerted efforts on the part of the breeder. Cross and Andersen (1994) recommended photographing nestlings to record the pin-feather development on their heads and then to use this as one criterion for selection of better head feathering—an excellent suggestion.

The Lutino colour morph, due to lack of melanin in the eyes, is subject to poor vision in bright light. Consideration should be given when housing Lutino specimens to choose aviaries and cages with some protection from glare, particularly in subtropical climates.

It has also been documented by avian veterinarians that Lutino specimens are prone to 'night frights'. In many instances this appears to relate to decreased vision and the bird not being able to find its way back to the perch after a fright. However, in some cases there are suggestions of seizure activity. These instances are not a direct action of the sex-linked Lutino gene as it is not a feature of this mutation in any other species of parrots. Instead it would be another associated fault bred into the colour strain. Therefore, it would be wise for breeders to select heavily against any bird exhibiting these traits.

Desirable Matings

The Lutino mutation combines best with the psittacin-altering mutations (Whiteface, Paleface, sex-linked Yellowcheek and dominant Yellowface) as well as the Opaline pattern gene (Pearl). One of the effects of the Opaline gene is enhancement of psittacin pigments and this is seen in the Lutino Opaline (Lutino Pearl) combination through extension of the yellow pigment within the plumage, particularly in a pattern across the top of the wings correlating to the pearl markings of a normal Grey Opaline.

Although it is generally expected that Lutino will mask all other melanin-altering mutations, a special gene interaction occurs with the Cinnamon mutation. This results in a bird that is referred to as a Cinnamon-ino, which is almost white in colour, but with a pale fawn suffusion throughout the plumage. The lutino and cinnamon loci lie close together on the Z chromosome with a recombinant frequency of 3%. This means that crossover is required to combine the two mutations and once it occurs, the tight linkage will result in the combination being further inherited as a single unit, unless another rare crossover occurs. In Budgerigars the Cinnamon-ino combination is known as 'Lacewing' and this name is sometimes used in other species of parrots as well, even though there is no 'lace pattern' in most species which lack wing markings like a Budgerigar. In Cockatiels, the term 'Lacewing' has unfortunately been used for variations of the Opaline (Pearl) mutation and this can become confusing for breeders who understand that a 'real Lacewing' is simply a Cinnamon-ino.

The only other melanin-altering mutation that can be combined productively with Lutino is the Australian gene known as Platinum and this is a special case because Platinum is an allele of the sex-linked ino locus. This relationship is explained in more detail in the *Platinum* section on page 139.

Combining the melanin-altering Pied gene with Lutino would seem an unlikely combination, however the anti-dimorphism effects of the Recessive Pied mutation results in increased yellow pigmentation throughout the plumage. This then becomes clearly visible in combination with the Lutino mutation. Adding the Opaline gene to produce a Lutino Pearl Pied produces specimens with the greatest possible yellow colouration in Cockatiels.

The combination of Whiteface and Lutino produces the Albino colour morph. Occasionally you might still hear a breeder claim that these 'are not the true Albino' and that 'one day a single mutation will occur to produce an Albino'. Unfortunately these breeders do not understand that the genetic control of melanin and psittacin pigments are separate pathways and therefore it will always require two mutations to block the two pigments. While it is genetically correct to refer to these birds as Whiteface ino, the term Albino is widely understood and equally acceptable to describe the colour, as long as the breeder always remembers that it is a combination colour and *not* a mutation.

Cinnamon (Z^{cin})

The Cinnamon mutation is another sex-linked recessive colour morph that is common in Cockatiels, being first bred in Belgium in 1968 (Sindel & Lynn 1989). George Smith (1978) reports that in 1968 Mr van Otterdijk discovered that an unnamed Belgium breeder had already established the Cinnamon mutation, indicating that the first bird was bred many years prior to this date.

The action of the Cinnamon gene is to prevent the final stage of eumelanin production, which results in the bird being unable to produce black pigment. As a consequence all melanin is changed from shades of black and grey into various shades of brown. This makes the Cinnamon Cockatiel an even brown tone throughout, with no alteration to the psittacin pigments in the plumage.

It is important to understand that the gene is *not removing grey* from the plumage, which would leave fewer melanin granules present. It is causing an 'abnormal' brown melanin granule to be produced. This is termed a 'qualitative' mutation because it

Cinnamon Cockatiel cock.

is altering the quality of melanin deposited in the feathers, not the quantity of melanin.

All melanin in the body is altered, which explains why Cinnamon birds are described as having 'plum'-coloured eyes when hatched, which become brown after a week or so. The melanin in the beak and legs is also altered becoming a pale beige colour. This feature of generalised melanin alteration in soft tissue structures as well as plumage is why Cinnamon is classified as a form of albinism. *Note: the category term 'albinism' is distinct from the colour term 'Albino'.*

The Cinnamon mutation became available to the general public in Australia by the start of the 1990s. This colour morph may have appeared independently in different locations, because the Cinnamon Cockatiels imported legally into Australia from the UK during the mid 1990s appeared to be a darker tone than the established Australian strain of Cinnamon. Yet interbreeding between the two strains confirmed that they were the same mutation. The Australian strain is reported to have originated from a wild bird caught near Kalgoorlie, Western Australia and Jeanette Hickford is given credit for establishing the strain in 1984 (Cross & Andersen 1994) or 1985 (Sindel & Lynn 1989).

The cinnamon locus in birds is directly equivalent to the autosomal brown locus of mammals which codes for tyrosinase-related protein-1 (TRP-1). This might seem a strange statement, to correlate an autosomal mammalian locus with a sex-linked avian locus, until it is understood that the X chromosome of the mammal is unrelated to the Z chromosome of the bird, with the two chromosomes having few genes in common. It makes even more sense when we learn that the avian Z chromosome carries many of the genes found on the mammalian chromosome 9 and that TRP-1 is found on this chromosome in mammals. (It should be noted that previously the cinnamon locus was theorised to code for TRP-2. However this locus would produce a slaty-grey rather than brown phenotype.)

As mentioned previously, the Cinnamon gene has a close relationship with the sex-linked Lutino gene. Firstly, the two loci sit close together, resulting in a small recombinant frequency of 3%. Secondly, the two genes interact to produce an unexpected phenotype in the Cinnamon-ino bird.

Cinnamon Cockatiels are generally strong healthy birds and there are no deleterious effects associated with Cinnamon strains. It is, however, well documented that cinnamon eumelanin granules are more susceptible to fading compared to normal black eumelanin granules, when exposed to extended periods of strong sunlight. This trait is not commonly observed in the Northern Hemisphere but is well known in Australia and other Southern Hemisphere countries from lower latitudes. Breeders living in those regions often have to house Cinnamon birds in well-sheltered aviaries to ensure their best plumage for the purpose of exhibitions.

Desirable Matings

Cinnamon combines well with the psittacin-altering mutations (Whiteface, Paleface, dominant Yellowface and sex-linked Yellowcheek) and the Opaline (Pearl) pattern gene. Because Cinnamon merely alters melanin colour while retaining good depth of tone, it can also be combined with a number of melanin-altering mutations to produce further colours. Cinnamon Pied is perhaps the most commonly bred, with other combinations being rare. In theory, Cinnamon could produce distinctive combinations with Platinum, Faded, Dominant Edged (Dominant Silver), Edged Dilute ('Silver Spangle') and Dilute ('Pastel' Silver) but few have been investigated. Combinations with the various Fallow mutations, 'Emerald' and Suffused (Australian 'Olive') are likely to result in indistinct colours that are virtually impossible to identify.

Cinnamon Cockatiel hen.

As discussed in the previous section on sex-linked Lutino, its combination with Cinnamon produces the unexpected Cinnamon-ino colour morph. Cross and Andersen (1994) credit Jeanette Hickford as the first to produce this combination in Australia in 1984. Due to the close genetic locations of the two genes, once combined the Cinnamon-ino tends to inherit as a single entity and is very easy to reproduce.

Cross and Andersen (1994) also believe that 'Lutino Cinnamon is much more striking when combined with the pattern changes of Pearl and Pied. Yellow pearled feathers are edged with cinnamon across the backs and shoulders. Lutino Cinnamon Pearl, Lutino Cinnamon Pied and Lutino Cinnamon Pearl Pied are startlingly beautiful birds when seen at maturity.'

In Australia, both the Platinum and the Cinnamon mutation became available to the general public around the same time. This resulted in dealers, unfamiliar with the new colours, selling mixed pairs of the two mutations, leaving many breeders disappointed when they produced Normal (Grey) offspring (the sons). As a result, many double split birds entered the gene pool, which caused a degree of confusion about the genetics of the two mutations for a few years. Fortunately, due to the tight gene linkage between the two mutations, no long-term damage was done to the development of either colour morph.

Recessive Pied (r)

The Recessive Pied mutation is the oldest of the established colour morphs in Cockatiels. Sindel and Lynn (1989) report that they were first bred by Mr D Putman in the USA in 1951, however Alderton (1989) and Cross and Andersen (1994) report them as appearing in 1949. George Smith (1978) also reports that they appeared before 1951, which was the date when Mr Hubbell took over the development of the colour morph. Around the same time Mrs R Kersh was also establishing Pied specimens in the USA and these were the source of birds imported into Europe (Smith 1978). They became available to the general public in Australia during the late 1970s around the same time as the Lutino and the Pearl. It is now readily available worldwide. Early specimens often only carried small areas of pied markings, however breeders have been able to selectively 'improve' the colour so that today a good Pied specimen has melanin pigment restricted to the 'saddleback' pattern and is symmetrical.

Grey Pied Cockatiel cock.

Grey Pied Cockatiel hen.

Grey Pied Cockatiel hen.

Grey Pied Cockatiel. The anti-dimorphism effect of increased yellow pigments, towards levels present in hens, is clearly visible in this cock.

'Pied' is an avicultural term derived from 'Piebald' which is used in mammalian species. It is used to denote mutations that produce a broken pattern of normal pigmentation and melanin loss. The areas of melanin loss are referred to as 'pied areas' and correspond to an absence of melanocytes (melanin-producing cells) from these regions. A noted feature of Pied mutations is that pied markings have an increased likelihood of affecting the extremities. This means that the primary flight feathers, tail feathers and the legs and feet are commonly affected and lose melanin pigmentation. In contrast, the eyes remain the appropriate colour for any mutation that the Pied gene is combined with.

There are many different Pied genes known in bird species, with the common one in Cockatiels being a 'typical' Recessive Pied. All Pied mutations fall within a subcategory of melanin-altering mutations known as leucism. Technically, since leucism only alters melanocyte functions, psittacin pigments should be unaffected. However this does not seem to be the case in the Cockatiel, as Pied Cockatiels generally have increased yellow pigment through their plumage, although the orange of the cheekpatches is never altered. There is an explanation for this and it is quite simple yet very intriguing.

Edged Dilute Pied Cockatiel cock. Note the anti-dimorphism feature of melanin in the face of this bird.

P. ODEKERKEN

It has always been difficult to identify the sex of Pied Cockatiels, which is surprising for a species with strong sexual dimorphism. In 1999, through observations made of Pied mutations in other parrot species, it was suddenly realised that the common Recessive Pied mutations in parrots had a neutralising effect on normal sexual dimorphism (Martin 1999a). It is now believed that all common Recessive Pied genes in parrots restrict the male sexual plumage dimorphism. This can be as simple as loss of the blue cere in Budgerigars, the orange vent spot in Elegant Parrots or the neck ring in Indian Ringnecked Parrots. However, in other species with more extensive sexual dimorphism the changes are even greater. Pied Red-rumped Parrot cocks lose their distinctive red rump colour and most of their bright structural colour, making them the drab olive colour of the hen. All this happens irrespective of where the pied markings appear.

The Recessive Pied Cockatiel also has multiple colour changes. The gene increases yellow psittacin in the cock (a hen trait) and also increases melanin deposition in the face area (another hen trait). However, the presence of pied markings in different areas can limit the appearance of these features. For instance a large pied marking across the face will remove melanin from this area, so that the anti-dimorphic trait cannot be seen. Yet this type of feature will also remove the melanin pigment from a hen's face and make a hen look more like a cock. The effect of increased yellow psittacin is often made more obvious by the pied markings removing melanin from large regions of plumage on the wings, tail and body.

Debate rages from time to time as to whether this Pied mutation is best described as recessive or as dominant. The conflict occurs because it is common for Normal split Pied birds to show the odd small pied feather on the back of the neck or pied feet. Some authors argue that this indicates a dominant trait. However, if this was a dominant Pied, then both single factor and double factor specimens would produce a significant pied pattern with no reliable distinction between the two genotypes. It is then suggested that this is a co-dominant trait, but if this was the case, then the single factor specimens should produce roughly half the pattern of the double factor specimens (midway between mutant and wildtype). Instead the heterozygous birds only ever show a few percent of pied markings, irrespective of how great the pied pattern of their parent was.

This mutation is indeed a Recessive Pied and the feature of 'visible split' is common in all species of birds with Recessive Pieds. The anti-dimorphism action of the gene also confirms the identity of this mutation, as no known Dominant Pied mutation exhibits this trait.

The facts are that Pied inheritance is far more complex than we would like it to be, so that it could fit into a simple category. If we take a broad look at Pied-type mutations in all species of animals (including Piebald and Spotting genes in mammals) it soon becomes apparent that there are about 10–20 genes involved in producing these patterns. However many of the genes produce only small effects when present on their own. To simplify the picture we can group these genes into the following categories:

Major Pied genes involve at least one major Recessive Pied and perhaps two major Dominant Pied genes. These produce a significant pied pattern on their own, which can be selectively increased.

Minor Dominant Pied genes produce only a small degree of pattern (less than 10% pied pattern). At least five different genes are potentially involved.

Minor Recessive Pied genes produce only a small degree of pattern (less than 10% pied pattern). At least another five different genes are potentially involved.

Modifier genes do not produce a pattern on their own (if no Pied genes are present) but can increase the pattern of other Pied genes if carried together. Yet another five different genes are potentially involved.

It is now apparent that most of these genes interact with one another, which is why all Pied mutations in every species are always capable of being selectively increased in pattern. One of the major Dominant Pied genes interacts with the major Recessive Pied to produce a Clear Pied phenotype, the other one does not. However, virtually all the minor Pied and modifier genes will produce varying increases in pied pattern in both major Recessive and major Dominant Pied mutations.

This means that when a well-marked Recessive Pied, as we have in the Cockatiel, is mated to a pure wildtype specimen, the likelihood is high that the Pied will also carry a varying number of minor dominant Pied genes as well as minor recessive Pied genes and modifiers. And, indeed, a split Pied showing a degree of markings will be carrying more of these genes and will potentially produce 'better' Pied offspring when mated back to a Pied bird compared to a split Pied not showing any markings. However, split Pieds should not be further outcrossed through non-Pied strains as future appearance of pied markings will not necessarily be indicative of the presence of a major Recessive Pied gene still being carried by the bird. The minor genes and modifiers can easily separate out from the main gene and continue to 'destroy' the purity of non-Pied colours, yet not necessarily produce attractive Pieds either.

The lesson to learn from this complex information is that the basic mutation must be treated as a recessive gene. Although the visual splits can be better to produce improved colour in future generations, they must be bred back to Pied specimens to ensure continued production of good pied features. It is also important to maintain other mutations in strains totally free from Pied as it is very difficult to ever totally eliminate Pied traits once they are introduced into a breeding line. It might be possible to select against a single recessive trait, but almost impossible to select against 10 different minor genes all at once.

When the Pied mutation first appeared in Australia, birds generally had smaller degrees of pied markings. Today, breeders have used selection mechanisms to produce the full 'clear' Pied. This is a bird that is genetically a Recessive Pied and carries enough minor genes and modifiers to result in a completely 'clear' phenotype. Potentially, within the make-up of these birds, there could be a significant minor Dominant Pied, the equivalent to Dominant Pieds present in other species that are used to produce 'Black-eyed Clear' combinations (eg Budgerigars and Red-fronted Kakarikis).

In those species, breeders have learnt to discern small differences in pattern between Dominant Pied and Recessive Pied. This then allowed the establishment of different breeding strains and their identification as distinct mutations. Yet so far in Cockatiels no-one has established a true Dominant Pied mutation—an area of potential development yet to be explored.

There is also potential for a single gene Clear mutation. This gene has been discovered in a number of species, including Red-rumped Parrots and Peach-face Lovebirds, so could be expected to appear in Cockatiels as well. It is a recessive gene that produces a greater than 95% clear phenotype every time. Many Europeans prefer to call this another Recessive Pied mutation, but I believe that it is distinctive enough to have its own identity. Besides which it can never be selected for a typical Pied phenotype of broken melanin pattern. When this mutation appears, breeding 'clear' forms of the standard Recessive Pied can lead to confusion with the new colour.

The 'Clear' Pied Cockatiel has been bred from the Recessive Pied mutation via selection.

One other potential mutation that has yet to appear in Cockatiels is the Mottled. This is a progressive leucistic mutation, roughly similar to someone's hair turning 'white' as they age. This mutation begins life as a normal-looking bird and then, with subsequent moults through its life, it gradually gains increasing mottled 'pied' areas in its plumage. Eventually some specimens can be almost entirely clear. To identify this mutation, breeders need to observe closely any Pied morphs for patterns changing over time. Standard Pied mutations generally have set patterns that alter very little from year to year.

Desirable Matings

The Pied mutation combines well with any other melanin-altering mutation that retains sufficient melanin pigment to produce a good contrast between pied markings and coloured regions. I would question the value of combining Pied with very light colour morphs such as the Fallow, Lutino and even Platinum, although some breeders like colours like Platinum Pied and other subtle combinations. My view is that the best Pied colours require excellent contrast and a good Whiteface Grey Pied cannot be beaten for this impact.

Apart from Whiteface, the other psittacin-altering mutations (Paleface, dominant Yellowface and sex-linked Yellowcheek) all produce attractive Pied combinations. The Opaline pattern gene can also produce attractive Pearl Pieds, as long as sufficient melanin and pearl pattern is retained. Unfortunately many Pearl Pied combinations do not carry enough melanin. The Opaline gene was combined with the Recessive Pied soon after the appearance of Pearl birds in West Germany in the late 1960s (Cross & Andersen 1994). This would probably have been the first combination colour produced in Cockatiels.

Whiteface (b)

The Whiteface colour morph is reported by Sindel and Lynn (1989) as appearing in the Netherlands in 1976, whereas Alderton (1989) gives an earlier date of 1969, with birds being available in Germany and the UK by the late 1970s. They appeared in Australia by the end of the 1980s and became available to the general public by the early 1990s. It is now one of the most popular colours in Cockatiels and is readily available worldwide. In some early publications, this mutation was named 'Charcoal', particularly for hen specimens, but nowadays it is universally known as Whiteface.

The term Whiteface is acceptable as the colour name for this mutation but it is important that breeders recognise that the gene mutation is correctly called Blue. Blue mutations occur in many species of parrots, with the basic action of the gene being the loss of all psittacin pigments from the plumage. All Blue mutations are inherited in an autosomal recessive manner.

Whiteface Grey Cockatiel cock.

Since the Cockatiel cannot produce structural colour, no visual blue colouration is evident. Instead, the loss of yellow and orange from the plumage produces a basic grey bird with a white face. In the hen, the face area retains the female trait of increased melanin. In addition, the sexually dimorphic pattern evident under the flight and tail feathers of hens, which normally appears yellow, is retained as a white pattern. Therefore, despite the loss of psittacin pigment, the colour morph is still easily sexed.

The psittacin pigments are also absent from the down of chicks, making the colour morph easily identified by white down instead of the normal yellow down. Melanin pigmentation both within the plumage and body tissues is unaffected, which means that the eyes, beak, legs and feet are dark or as per any combined mutation.

The exact mode of action for the blue locus has not been investigated by science, therefore we do not know whether the gene is involved in production, distribution or deposition of the psittacin pigments. All we can deduce is that the blue locus is critical for a very important step in the use of psittacofulvin pigment in the plumage of parrots.

Cross and Andersen (1994) report that during the establishment of the Whiteface mutation in Australia 'it suffered as a result of inbreeding to produce enough stock to develop the strain. For those breeders able to obtain birds, the general complaints were that some of the birds were small, poor in quality and nervous. They were difficult to pair for breeding and produced a high

Whiteface Grey Cockatiel hen.

incidence of infertile eggs. Subsequent generations have improved steadily in fertility and stability.'

Desirable Matings

The Whiteface is a striking bird of contrast and combines well with mutations altering differing aspects of the colouration—the melanin-altering mutations and the Opaline (Pearl) pattern gene. Combinations such as Whiteface Cinnamon, Whiteface Lutino (Albino), Whiteface Pied and the various 'Silvers' all produce distinctive phenotypes that are very attractive.

In fact, combining a melanin-altering mutation with Whiteface is informative regarding the nature of the melanin-altering mutation. Since Whiteface only removes yellow and orange pigments from the plumage, we are given a much improved view of the melanin changes present in the other component of the combination. Yellow pigments are notorious for confusing the human eye about differing shades between grey and brown. With no yellow remaining in the Whiteface combinations, the true tone of melanin colour can be observed clearly. When comparing and evaluating different melanin-altering mutations, the Whiteface combination is an important bird to consider.

Whiteface Grey Cockatiel cock. The contrast produced by the Blue mutation is always eye-catching.

Whiteface Grey Cockatiel hen.

It is not wise to combine Whiteface with psittacin-altering mutations as the total lack of psittacin pigments will mask the presence of either dominant Yellowface or sex-linked Yellowcheek. With the Paleface ('Pastelface') mutation the situation is a little different because this gene is an allele of the blue locus. As a result it is common practice for breeders of the Paleface colour morph to always mate their birds to Whiteface, which means that virtually all 'Pastelface' birds are actually heterozygous PalefaceWhiteface genotypes. In fact I doubt whether too many breeders have ever even seen a pure Paleface bird. I will discuss this special genetic relationship further in the Paleface section.

The combination of Whiteface and Lutino produces the Albino colour morph. Many authors maintain that somehow a single gene Albino will one day appear. However this is impossible as melanin and psittacin pigment production and usage each have completely independent genetic control. It is therefore impossible for a single gene to ever block both pigment types. In all parrot species that carry both pigment types, an Albino can only ever be a combination colour involving both Lutino and Blue genes. It is therefore entirely appropriate to call this colour Albino, but it should not be referred to as a 'mutation' as it is the result of a combination of mutations.

Paleface ('Pastelface') (b^{aq})

Neither Sindel and Lynn (1989), Alderton (1989) nor Cooke and Cooke (1993) give any information on this mutation, yet it is listed by Jim Hayward (1992) writing from the UK. Therefore, I conclude that it must have appeared around the start of the 1990s. Indeed the mutation was legally imported into Australia in the mid 1990s so it must have been well established by that time and is now found worldwide in significant numbers.

This mutation is universally known as 'Pastelface' and I fear that it is too late for breeders to be able to accept a name change. Unfortunately this is an inappropriate name as the term 'Pastel' implies a melanin-reducing trait and this mutation is in fact a Parblue gene altering psittacin pigments. The error in the initial use of this name in the UK, Australia and a few other countries stems from the incorrect use of 'Pastel' for Parblue mutations, eg 'Pastel Blue' in Peach-face Lovebirds and Indian Ringnecked Parrots. When the Parblue gene appeared in Cockatiels, breeders borrowed from these other species and called the colour 'Pastelface'. I acknowledge that breeders are reluctant to adopt new names, but it would be advisable to do so now, before a true Pastel mutation appears in this species, which could lead to significantly more confusion.

Paleface Grey Cockatiel cock.

Paleface Grey Cockatiel cock.

Paleface Grey Cockatiel hen.

Like all Parblue mutations, this gene is literally a partial blue gene. This means that it behaves as a partial copy of a Blue mutation, reducing the ability to use psittacin pigments in the plumage to a partial degree. As a result, the yellow of the face becomes pale, the orange cheekpatches are much paler but remain an orange shade and all other yellow pigments (in the hen) are also reduced in intensity. Overall, the reduction in psittacin pigment is roughly 50% of normal and appears to be even throughout. This would indicate that this particular Parblue mutation would be categorised as an Aqua gene.

Being termed a Parblue mutation also means that Paleface is a multiple allele of the blue locus. This latter genetic point implies that, while the Paleface gene itself is autosomal recessive, it resides at the same position as the Whiteface gene and the two genes behave co-dominantly towards one another. The heterozygous PalefaceWhiteface bird is phenotypically also Parblue, but is a different colour from a homozygous Paleface bird. Due to the general lack of psittacin pigments in the Cockatiel, it would be difficult to quantify the difference visually between the homozygous and heterozygous birds.

Some breeders have begun calling the two different forms as 'Single factor' and 'Double factor'. However, this terminology should be reserved for true dominant mutations, as the term 'Single factor' implies one mutant gene and one wildtype gene, not two different mutant genes. Other breeders refer to the birds as being 'split Whiteface', but once again this is inaccurate since the Whiteface gene is contributing exactly half of the effect on the phenotype. In other words, the appearance of the bird is due equally to both the aqua (Paleface) gene and the blue (Whiteface) gene. As such, using the blended-name terminology is the most appropriate way to describe the colour morph.

Borrowing from the diagrams used by Diana Andersen in the first edition of this book can help illustrate how and why Blue and Parblue genes interact. Each is a different mutation of the same gene, therefore they can only occupy the same space (loci) on their chromosome. This means that a bird can only have a maximum of two genes and therefore only the following genotypes are possible with respect to these two mutations.

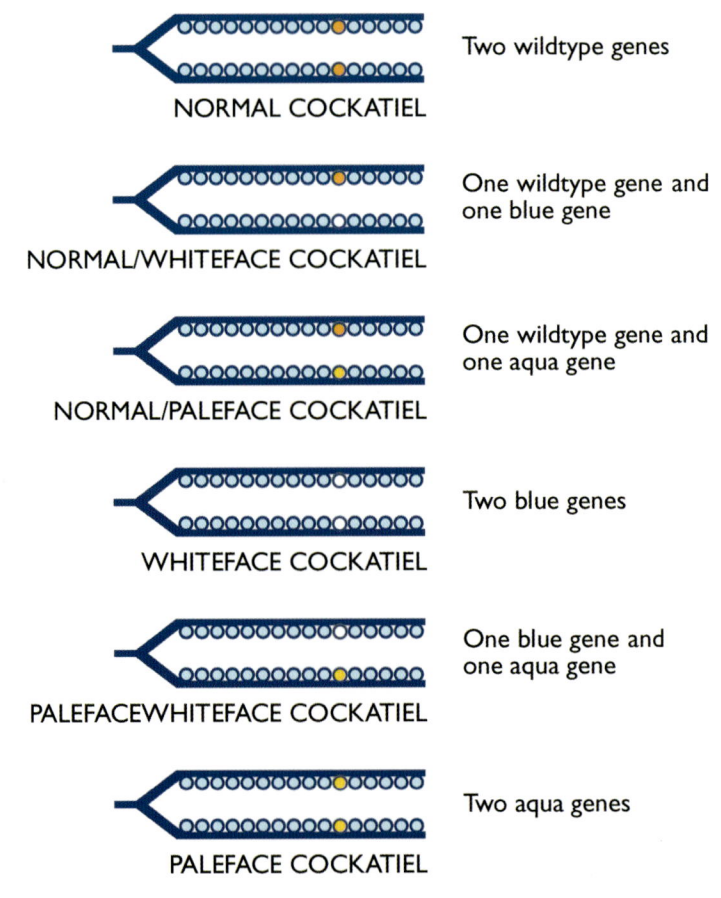

NORMAL COCKATIEL — Two wildtype genes

NORMAL/WHITEFACE COCKATIEL — One wildtype gene and one blue gene

NORMAL/PALEFACE COCKATIEL — One wildtype gene and one aqua gene

WHITEFACE COCKATIEL — Two blue genes

PALEFACEWHITEFACE COCKATIEL — One blue gene and one aqua gene

PALEFACE COCKATIEL — Two aqua genes

One area of potential for new colours in Cockatiels would be the identification of new Parblue mutations. We know already from other parrot species that the blue locus can mutate in many ways and produce distinctly different colour morphs. Potentially there may be at least three or four different Parblue mutations as well as the Blue gene.

The difficulty for Cockatiel breeders is the overall small degrees of psittacin pigmentation in this species. This leaves little room for identification of variations in Parblue phenotypes.

Desirable Matings

Paleface combines well with all mutations of other classes, either melanin-altering or pattern genes. In fact, one useful way of dividing Cockatiel colours up is to separate them by 'face colour'. This method groups colours into Normals, Paleface combinations, Whiteface combinations, Yellowface combinations and Yellowcheek combinations.

As explained already, being an allele of the blue locus, it can be mated to Whiteface without concern, but caution should be used in combining it with other psittacin-altering mutations because the combined birds will be difficult to distinguish. In general, it would be wise to keep Paleface separate from either dominant Yellowface or sex-linked Yellowcheek breeding lines. If they were combined together, the resultant birds would have a pale yellow face and cheekpatches, overall an insipid result compared to the base mutations.

Opaline (Pearl) (Z^{op})

The Pearl colour morph was first bred in Germany in 1967, with that date agreed to by both Sindel and Lynn (1989) and Alderton (1989). It became available to the general public in Australia in the late 1970s at the same time as the Lutino and Pied mutations appeared. It is now one of the most common mutations worldwide. The Pearl pattern is now recognised as the action of the Opaline mutation in this species.

Stan Sindel (1989) disagrees with this conclusion. He believes that the pattern is produced as an 'extension of the pale yellow barring and spotting present in the adult female… which accounts for the young males moulting out the Pearl plumage on the adult moult'.

I agree with his observation but not his conclusion. The yellow spotting on the Cockatiel hen is directly analogous to the underwing stripe of other parrot species and it is a common feature of many Opaline mutations that these features are enhanced. Therefore Sindel's observation does not rule out Opaline, but instead supports its classification as the gene responsible for the Pearl colour.

Opaline is a difficult mutation to quantify in its action, in fact we do not really know how it plays its role. What we can say is that the wildtype gene for the opaline locus is involved somehow in the normal production of the plumage pattern. When it mutates, it produces a range of alterations to the plumage pattern that varies with each species in which it occurs. The changes that we see are best described as 'general trends' and because the plumage pattern is different in each parrot species, the interactions between Opaline and other specific genes determining each species' unique appearance will be complex and produce some degree of variation in different Opaline morphs.

There are six basic changes seen in Opaline colour morphs across species that help identify them as Opaline mutations (Hesford 1998; Martin 1999b, 1999c). Not all species show all six effects, but taken as a large group of mutations across species, the commonality becomes more obvious. The six traits are:

- sex-linked recessive inheritance
- pattern-altering gene
- altered melanin pigmentation
- enhanced psittacin pigmentation
- enhanced underwing stripe
- loss of melanin from natal down

Opaline Cockatiel cock. Note that the pearling has almost moulted out with maturity.

The first trait is simple and obvious. A colour morph could not be considered an Opaline if it is not inherited in this manner. Its locus would have a different function from the one found on the Z chromosome, even if there happened to be great similarities in appearance.

Opaline is currently the only recognised mutation in the category best referred to as 'Pattern-altering genes'. In my first book, **A Guide to Colour Mutations and Genetics in Parrots**, I used the term 'distribution gene' for this category. Since that time, I have been made aware that this terminology could be confusing for some breeders and particularly scientists who relate the word 'distribution' with the actual process of moving pigment within the skin or plumage. This was not my meaning when I defined this feature of Opaline mutations and I now believe that the term 'Pattern gene' is a far more accurate terminology.

While all other mutations alter plumage pigmentation by interfering in some step in the process of either melanin or psittacin production or deposition, the Opaline gene has no effect on the actual pigments produced—it simply alters their pattern throughout the plumage. As the only recognised Pattern gene, it currently is the only known gene that directly alters both pigment classes within the plumage. Note: *the anti-dimorphic effects of Recessive Pied mutations can alter both pigment classes indirectly by changing the bird's sexual dimorphic traits.*

The Pearl Cockatiel is produced by the Opaline mutation.

The third trait of altered melanin pigmentation is complex and varies slightly for each species. The melanin pigmentation will remain normal in many areas of the plumage, particularly the flight feathers. However, in general terms, there is often a reduction of foreground melanin deposits, while background melanin can either be reduced or increased. In some species, this latter change can actually increase structure colouration in the plumage (such as green replacing yellow colouration across the back of Opaline Budgerigars). The important feature of the melanin changes is that wherever melanin is produced, it is produced in normal quantities and quality, thereby ruling Opaline out of the albinistic and dilute categories.

Enhanced psittacin pigmentation is an uncommon colour change for mutations in parrots. Yellow pigment might be increased in both quantity and distribution through the plumage pattern. In some species like the Budgerigar, yellow is already evenly strong through the plumage, and therefore cannot be increased any more. However in a species like the Cockatiel, where yellow is normally quite restricted, there is plenty of scope for increase. The increase in yellow pigmentation within the plumage is even more noticeable when Opaline is combined with strong melanin-reducing genes such as Lutino, 'Emerald' or the various Fallow mutations. In species that are able to produce red or pink psittacin pigments, these are also increased. However, they are never added to the plumage of a species that does not produce them somewhere in the wildtype. Therefore there is no red or pink in the Pearl Cockatiel.

In parrots, the only other colour morphs that show increased psittacin pigmentation are known as 'suffusion' traits. These colours behave as selective traits that increase a particular psittacin pigment throughout regions of the plumage. Examples include the Red-fronted mutations in Scarlet-chested and Turquoisine Parrots. In the Cockatiel there are suffusion genes that increase yellow psittacin throughout the plumage and these have been well established in certain strains. Yellow Suffusion should not be confused with the Suffused mutation, both of which are discussed in more detail further on in the book.

The fifth Opaline trait is the enhancement of the underwing stripe found in many

Opaline Cockatiel hen.

wildtype hens of different species, particularly Australian parrots. This trait is not found in all Opaline colour morphs, for instance the Peach-face Lovebird and the Regent Parrot (Sindel 2003). In the Cockatiel, the wildtype hen has a series of underwing spots rather than a true stripe as seen in many other species. As discussed already, the pearl markings of the Opaline hen are clearly an enhancement of these spots across the entire plumage. While it is true that in the adult Pearl Cockatiel cock the pearl markings are lost, Cross and Andersen (1994) explain this effectively: 'If he is lightly pearled, the increase of melanin as he approaches maturity more fully masks the markings. Cockatiel cocks do not lose their pearls, their markings are simply covered over by the necessary melanin that indicates sexual maturity. When Pearl cocks are aged or ill enough that they do not produce breeding hormones their pearls will become more visible again.'

Considering the range of variation seen with Opaline colour morphs across parrot species, this feature of the adult Pearl Cockatiel cock is not in conflict with the genetic classification of Opaline.

The last Opaline trait is loss of melanin from the natal down. The chicks of many species of parrots are grey due to the presence of melanin pigment. This is absent in specimens of Opaline mutations. However, the Cockatiel lacks melanin in the natal down to begin with and therefore this trait cannot be assessed. It has also been noted by many breeders that the down of Opaline specimens is slightly thicker, longer and lusher. However this assessment is very subjective and not consistently noted by all aviculturists.

Cross and Andersen (1994) describe the features of the Pearl Cockatiel nestling very precisely: 'In the nest the Pearl mutation cannot be identified until the chick is at pin-feather stage. The pin-feathers will look dotted or striped before the feathers break out of the feather casings. Its crest breaks through the skin a normal colour for the variety involved and its eyes are the colour that would be normal for the base colour that is combined with Pearl. After fledging, legs and feet are a pinkish tone with a grey overlay of colour. Some families have quite pink feet, but the toenails are always dark grey in a Normal (Grey) Pearl.'

Like all Opaline mutations in other species of parrots, the Pearl Cockatiel is variable and quite amenable to selection. This allows the breeder to selectively increase many of the effects of the opaline gene, including the extent of yellow pigmentation and size and shape of the pearl markings. In some countries there are standards written for different shaped pearls, while in others, breeders aim for a specific ideal. The name 'Lacewing' has been used in conjunction with certain selected forms of Opaline but should be avoided because the term correctly refers to the Cinnamon-ino combination in Budgerigars.

Desirable Matings

The Opaline mutation, being a pattern gene, readily combines with all other currently known mutations, many producing extremely attractive and desirable combination colours in Pearl. When combined with the more heavily reduced melanin-altering mutations (eg Lutino, Suffused or 'Emerald'), the enhancement of yellow pigments is particularly visible and attractive. In contrast, combinations with Normal (Grey) and the darker mutations such as Cinnamon, Dilute and Faded, are stunning when combined further with a psittacin-altering mutation such as Blue (Whiteface). When producing new combination colours, the Opaline gene is perhaps the most flexible mutation available in

Cockatiels, with only the Albino Pearl being indistinguishable.

The combination of the Opaline gene and the Recessive Pied gene is an interesting one because of the anti-dimorphism effects of the Pied mutation. These effects increase the retention of pearl markings in the adult Pearl Pied Cockatiel cock, as long as the pied markings are not too extensive and sufficient melanin is retained by the bird.

Cross and Andersen (1994) questioned the wisdom of combining Opaline with the Edged Dilute mutation ('Silver Spangle') and their point was valid to some degree. There exists conflict between the pearl patterning of the Opaline mutation and the melanin edging effect of the Edged Dilute. The combination can be produced but, in theory, it would be difficult to observe the distinguishing features of the Edged Dilute, thereby making the colour look more like other 'Silver Pearl' combinations.

A similar dilemma confronts overseas breeders combining the Dominant Edged mutation with Opaline. Photographs further on in the book show that it is possible, but that the edging trait is difficult to observe.

Cross and Andersen (1994) report that 'the pearled feather pattern often contributes better feather quality to other varieties when bred together. Even birds that do not display pearls, such as a Normal hen bred from a split Pearl father, may display a 'crowned' type of crest that is typical of Pearl birds. Using Pearl family lines can improve head feathering in any family.'

Yellow Suffusion

Cockatiels carrying the Yellow Suffusion gene are well established worldwide in many strains of birds. However it is often overlooked, or worse still, mistakenly considered an action of one of the more obvious melanin-altering mutations present in the bird. Yet breeders have always recognised that different strains of Cockatiels carry different degrees of yellow psittacin in their plumage and they have deliberately introduced it into the colours that they felt were more attractive with it (eg Pied), while selecting against it in other instances.

Recently one of the US Cockatiel societies has adopted the name 'Suffused Yellow' for the 'Emerald' mutation. They have made this decision in the belief that this mutation is a melanin-altering mutation known as Suffused. If this were true, the colour should be referred to as Suffused Grey (or Cinnamon or whatever the base colour was) as yellow pigmentation in this mutation is unchanged and simply more visible due to melanin reduction. I will discuss further in a later section why I believe that this name choice is incorrect and that the true Suffused mutation has appeared independently in Australia. At this point it is sufficient to realise that when US breeders speak of 'Suffused Yellow' they are not referring to the Yellow Suffusion mutation being discussed in this section.

Yellow Suffusion is a selective trait akin to Red Suffusion found in some other species of parrots (eg Red-fronted Scarlet-chested Parrots). However, being selective does not imply a lack of genetic control. Instead, it implies a more complex form of inheritance either involving multiple gene modifiers or a new concept known as 'tandem repeats'. If multiple gene modifiers are the true mode of inheritance, then strains with greater degrees of yellow colouration in the plumage carry a lot more independent modifier genes than a strain with little yellow. But if the tandem repeat concept is at work, a single gene for yellow suffusion can vary in the number of important repeated DNA sequences that it carries within its gene. The more repeats the gene carries, the greater the expression of yellow pigmentation.

In either situation, breeders are able to selectively alter a strain of bird to either carry more or less yellow pigmentation throughout the plumage. In the case of darker-coloured morphs, this trait may be largely unnoticed due to the heavy loads of melanin within the feathers. However it becomes very noticeable in strains of mutations that greatly reduce melanin and often leads to confusion for breeders who mistakenly believe that the varying degrees of yellow somehow relate to the melanin-altering gene (eg Lutino, Suffused or 'Emerald').

Finally, breeders should recognise that it is also possible to have increased yellow pigmentation due to an acquired non-genetic colour change. These birds are often believed to be suffering from liver disease and develop increasing yellow pigmentation through their plumage over time, in contrast to the genetic forms that remain stable in their degree of yellow colouration.

Desirable Matings

It is up to the breeder to decide if they prefer a particular colour with a greater or lesser degree of yellow suffusion. It does, however, complement the colour of many Pied specimens and also the Opaline gene. Of course it is negated by the Blue gene (Whiteface) and may lessen the appeal of Paleface birds by moving their colour back towards Normal. In theory, it would complement the colour of the Yellowface and Yellowcheek mutations.

One point that should be considered when deciding to select for or against Yellow Suffusion in strains of different colour morphs is the effect that yellow pigment has on our mind's interpretation of different colours. Yellow pigment is detected in our eyes by our 'red cone cells' along with red wavelengths of light. Brown colours have a red component while grey colours do not. Since yellow stimulates our red cone cells, Yellow Suffusion in a grey-shaded morph will make it look more brownish, while it will enhance the brown tones of mutations such as Cinnamon and Bronze Fallow.

As such it is probably more desirable to select against Yellow Suffusion in the various 'Silver' coloured mutations—Dominant Edged, Dilute, Platinum, Faded, Edged Dilute, Suffused and Ashen Fallow. Conversely, the effect of Yellow Suffusion can be viewed as positive in mutations such as Opaline and Pied which in themselves also enhance yellow pigmentation.

Silver

The name 'Silver' is mentioned briefly here because it is a name that is in widespread use all around the world, but unfortunately it is used for many different mutations, mostly incorrectly. It is, therefore, important for Cockatiel breeders to always clarify exactly which mutation is being discussed, particularly when discussing colours with breeders from different regions.

If used solely as a description of the physical colour of a mutation, the name 'Silver' either has or could be applied to eight different mutations around the world:
- Dominant Edged ('Dominant Silver')—Europe and USA
- Ashen Fallow ('Recessive Silver')—Europe and USA
- Dilute ('Pastel Silver', 'East Coast Silver')—Australia
- Platinum—Australia
- Faded ('West Coast Silver')—Australia
- Edged Dilute ('Silver Spangle')—Australia
- Suffused ('Olive')—Australia
- 'Emerald' ('Suffused Yellow')—USA

Of all these mutations, only the Dilute and the Suffused can truly be called Silver mutations. Silver should be correctly applied to Dilute Grey phenotypes, Suffused being a second form of dilution. However, it is my advice that breeders learn the correct genetic terms and use them from henceforth to aid better understanding and identification of the colours they breed.

At present, with so many colours being called 'Silver', it is common for breeders to mistakenly interbreed them, which then produces more and more birds with confused genetics. Eventually some of the rarer recessive genes could be lost in their pure form and their distinctive traits lost as more and more blended colours comprising multiple mutations gradually appear. I direct the reader to the appropriate sections on each of these different 'Silver' mutations.

Single factor Dominant Edged Cockatiel cock.

'Silver' can be applied to eight different international mutations.

Ashen Fallow Cockatiel hen.

Dilute Cockatiel cock.

Platinum Cockatiel cock.

Faded Cockatiel cock.

Edged Dilute Cockatiel cock.

Suffused Cockatiel cock.

'Emerald' Cockatiel cock.

REGIONAL MUTATIONS

These mutations are a mix of new colours and those that have their distribution limited to only one or two continents. Generally speaking, the European and North American mutations have spread between those two regions but not Australia, while Australian mutations so far have not appeared outside of Australia.

European Mutations

Dominant Edged (Dominant Silver) (E)

The Dominant Silver was discovered in a pet shop in the UK in 1979 (Alderton 1989). It is now widespread and common throughout Europe and North America, as well as much of the rest of the world except Australia. This mutation has now been identified in a number of different parrot species (eg Indian Ringnecked Parrots and Fischer's Lovebirds) where it has been assigned the genetic name Dominant Edged. Genetically it is an autosomal co-dominant mutation.

Although some authors use the term Dominant Dilute, it is currently not clear whether it is, in fact, a true Dilute mutation or rather a form of dark albinism. Some breeders report that the eyes of chicks are a shade less than dark for a few days after hatching, which implies albinism. And, of course, the term 'Silver' cannot easily be applied to those species which are basically green in colour—hence the term Dominant Edged was created.

I vary from the European standard in one way, in the genetic symbol that I ascribe to this mutation. In Europe they use 'Ed', which is the same symbol that they use for the recessive Edged Dilute mutation ('ed'). This would indicate that the two different mutations are alleles,

Single factor Dominant Edged Cockatiel cock. Generally known as 'Dominant Silver' or 'Dominant Dilute' by Cockatiel breeders, this mutation is correctly known as Dominant Edged in a number of other parrot species.

one a dominant allele and the other a recessive allele, which is not true. Hence I have chosen to use 'E' on its own for the Dominant Edged gene and therefore avoid any possible confusion.

Although the exact genetic action of this gene has not been determined, there is a mutation in domestic fowl that has great similarities and might

Young single factor Dominant Edged Cockatiel cocks showing variation in the depth of plumage colour.

Single factor Dominant Edged Cockatiel hens.

very well be the same gene. In fowl the mutation is known as Blue. In single factor form, when combined with a Black fowl to begin with, it produces a deep blue-grey colour with a distinctive 'edging' pattern, seen at its best effect in the Andalusian breed. The edging effect is a concentration of melanin, or more precisely an area of the feather that does not have as much melanin reduction as the rest of the feather. In fowl, the double-factored form produces a largely white phenotype, known as Splashed, with the odd spot of retained melanin in some feathers.

These effects are very similar to those seen in Cockatiels and other parrots with the Dominant Edged mutation. The single factor phenotype reduces melanin in the plumage to varying degrees, but always has a tendency to leave a pattern of more concentrated melanin creating the edging effect. The double factor phenotype, in truth, gives a more accurate portrayal of the gene action by reducing all melanin heavily. In the 'best' specimens it will be almost clear (devoid of melanin) but always has a tendency to retain some degree of melanin.

There is much variation in the single-factored Dominant Edged which allows breeders room for selection. In the USA, breeders have largely steered away from breeding double-factored birds because of a perception that they are weak. Indeed, since the single-factored Dominant Edged can be continually outcrossed, it is both easy to reproduce and always remains highly vigorous. In contrast, producing the double-factored Dominant Edged requires the mating of two mutants and until now perhaps the gene pool in the USA might have been too narrow to sustain this type of mating. The result has been that USA breeders have tended to select the single-factored birds towards a lighter colour, which they might not have done if more double-factored birds were bred.

There is also a tendency to select against the edging effect, in order to produce an even silver colouration. I see this as a shame, as it neglects one of the key features of the mutation and as true Dilute mutations (as we have here in Australia) become available around the world, distinctive features will be essential to maintain identity of the different colours.

Page 127

Desirable Matings

The single-factored Dominant Edged combines well with the psittacin-altering mutations (Whiteface, Paleface, Yellowface and Yellowcheek) as well as the Opaline pattern gene. Since single-factored Dominant Edged retains significant melanin, it also produces distinctive Pied combinations. In theory it should be able to combine with Cinnamon, yet to date I have not seen a photograph of this combination despite the two mutations being common in Europe and the USA for many years. Indeed many USA breeders believe that the two genes are incompatible. This raises interesting possibilities from the perspective of understanding genetics and pigmentation and I would love to hear more from any breeder who has investigated this combination.

Other melanin-altering gene combinations are undesirable as they would result in light, indistinct colours that would be difficult to identify. For the benefit of the species and for the rare mutations, it is important that Cockatiel breeders keep the gene pools of the various melanin-altering mutations pure and free of other genes from this category.

The double-factored Dominant Edged, being so greatly reduced in melanin, really only combines well with psittacin-altering genes. Pearl and Pied combinations have little contrast and tend to be indistinct, although the photograph above proves that Pearl can produce an attractive and unique phenotype.

Double factor Dominant Edged Pearl (left) and Double factor Dominant Edged Cockatiels.

Ashen Fallow ('Recessive Silver') (ash)

The European 'Recessive Silver' is actually a Fallow mutation since it retains red eyes as an adult and has a general reduction of plumage pigmentation. As there are many different mutations referred to as 'Silver' worldwide, it is important to adopt the correct genetic name to avoid confusion between this mutation and the true recessive Silver mutations found in Australia. The mutation was established in the 1960s in Europe (Alderton 1989; Cross & Andersen 1994), but like many Fallow mutations has remained difficult to reproduce in strong breeding strains and is still uncommon today. Indeed, the German Cockatiel Society has banned its exhibition due to its poor vigour.

Ashen Fallow Cockatiel.

It now seems that this autosomal recessive mutation survives in its strongest numbers in the USA, which is where most of the photographs for this book originated. Elsewhere, the majority of Cockatiel breeders appear to have changed to breeding the much easier Dominant Edged mutation. I would encourage any breeder who has specimens of this mutation to continue the difficult job of trying to improve its vigour.

Cross and Andersen (1994) made the following interesting comments regarding the Ashen Fallow (referred to as 'Recessive Silver'): 'Many references in avicultural literature state that early strains suffered from lethal factors and blindness. I was fortunate to talk to Dr G Th F Kaal of Holland who was involved with these 'blind' Silvers … The full story was that only a few birds had problems and after the first generations the Red-eyed Silver Cockatiels were quite normal in eyesight and health. I observed Red-eyed Silver and Fallow birds in the USA in 1984 and at that time they were as strong and fertile as the Normal Cockatiels.'

Previously, I had always believed that Fallow mutations were only weak due the inevitable inbreeding associated with the development of recessive traits. Recently, however, I have been studying similar pigmentation genes in mammalian species including humans where much more is known. From this I have learnt that there are a number of genes that will produce a phenotype which we would call Fallow in parrots. We already knew this much as there are many parrots with two different Fallow mutations and Budgerigars originally had three types of Fallow.

Ashen Fallow Cockatiel hen. Known in Europe and the USA as 'Recessive Silver', this mutation must be renamed to avoid confusion with the true recessive 'Silver' mutations found in Australia.

I have learnt that Fallow-type morphs are produced in humans by the tyrosinase (NSL ino) locus, the adaptor protein 3 (AP-3) locus, the P protein locus and the membrane-associated transporter protein (MATP) locus. Mutations of some of these loci only affect pigmentation. However, mutations of some of the other proteins affect many other systems apart from pigmentation and are known to affect the long-term health of individuals expressing these genes. It is therefore not surprising that many Fallow colour morphs in aviculture have proved difficult to reproduce and improve in vitality.

Yet despite this, we should persist in our efforts as there are also many Fallow morphs that are improved and become just as viable as the wildtype. And it is also important to keep rare mutations going as they help us gain a broader perspective of how pigmentation works across different species.

As a Fallow mutation, the chick hatches with a red eye which is retained throughout life. The melanin is reduced in quality from wildtype, becoming an even, light silver shade throughout the plumage. Psittacin pigments are unaffected and retain their normal distribution. Since the melanin pigment is significantly reduced, the degree of yellow suffusion in the plumage becomes far more noticeable than in the wildtype. However, this feature should not be overly interpreted. Breeders must decide if they prefer the yellow suffused birds or the clearer silver melanin colours when they are selecting the strain that they choose to breed with, as well as any Normal birds used as outcrosses.

Desirable Matings

Like other melanin-altering mutations, this colour combines best with psittacin-altering genes and the Opaline pattern gene. Although Ashen Fallow Pieds are distinct and worthwhile, combinations with other melanin-altering mutations are less desirable since they would produce pale, indistinct colours. The only reason for combining the various melanin-altering genes is pure genetic study and any offspring of such matings should not be allowed to mix into the general Cockatiel population.

Sex-linked Yellowcheek (Z^{yc})

The sex-linked recessive Yellowcheek mutation was first bred by Mr Bruno Rehm in Germany around the early 1990s. Unfortunately Mr Rehm no longer keeps birds and has no record of the exact date. However, it is recorded that the mutation was imported into the USA from Germany in 1992 by Elsie Burgin, Nancy Rocheleau and Dave Okura. Therefore it was clearly established by this time.

It is still not common in the USA, perhaps because it has had to compete with the more easily bred dominant Yellowface mutation. The photographs published in this book have been sourced primarily from Germany where it is widely bred. In the first edition of this book, Cross and Andersen reported upon some specimens of this mutation that appeared in Australia. Unfortunately those birds were not established and at present there have been no further recorded breedings of this mutation in this country.

To avoid confusion between this sex-linked gene and the similar dominant trait bred in the USA, I suggest that this mutation is continued to be called Yellowcheek and that the autosomal dominant mutation be subsequently referred to as Yellowface. This also brings the dominant mutation into alignment with the same mutation in other parrot species. No doubt many breeders will resist this suggestion since 'old habits die hard'.

The Yellowcheek mutation is so far unique to Cockatiels. No other sex-linked psittacin-altering mutation has appeared in any other species of parrot. In my book **A Guide to Colour Mutations and Genetics in Parrots** I theorised that perhaps this mutation represents a gene specific to Cockatiels, perhaps intricately involved with the production of the cheekpatch. This was based on studying limited photographs of the mutation which suggested that the cheekpatch had merely

Yellowcheek Grey Cockatiel cock.

Yellowcheek Grey Cockatiel hen (left) and cock.

Above: Yellowcheek Grey Cockatiel hen. This sex-linked mutation is unique to Cockatiels.
Below: Yellowcheek Grey hen (left) and Normal Grey hen.

been lost from the face area. However, now that I have studied a larger number of photographs of the Yellowcheek mutation, it is clear that a defined cheekpatch remains. We are therefore left to wonder how this gene might act somewhere in the production pathway for orange psittacofulvins. And if this is the case, why do we not find this mutation in other parrot species that produce orange or red psittacin pigments?

The Yellowcheek produces a 'purer' yellow cheek colour than the dominant Yellowface gene and the colour does not develop the orangey tone as the cocks mature. Being a psittacin-altering mutation, it is ideal for combining with mutations from other classes and in effect creates a new 'series' of Yellowcheek colours.

Desirable Matings

All melanin-altering mutations combine superbly with the Yellowcheek, as does the Opaline pattern gene. It is, however, strongly advised that breeders avoid contaminating Yellowcheek strains with either Whiteface or Paleface genes. Whiteface would mask the presence of the Yellowcheek mutation and Paleface will produce weaker psittacin colours, losing the appeal of the strong yellow cheeks. Combining the Yellowcheek and Yellowface mutations would be interesting purely from a scientific point of view and indiscriminate matings could lead to the loss of one of these two mutations as pure breeding strains. Subsequent breeding with mixed strains would cause great confusion for breeders regarding genetic results and identification as the two different mutations separate out.

Non Sex-linked Lutino (a)

The autosomal recessive (NSL) Lutino mutation has been bred in Europe but remains in very low numbers. In appearance, it is identical to a sex-linked Lutino and since the latter mutation is so widespread, there is little chance that the NSL Lutino will ever become widely established. Its main value is in test mating other melanin-altering mutations to help identify their gene locus.

This has been done with the Bronze Fallow, where matings between the two mutations produce intermediate phenotypes (half way between Fallow and Lutino), instead of Normal double splits as would occur if the two mutations were not alleles. This is a parallel genetic relationship to that seen with the blue locus alleles (Whiteface and Paleface) and the sex-linked ino locus (Platinum and Lutino).

Any Cockatiel breeder who has specimens of the NSL Lutino should be encouraged to perpetuate the mutation because of its importance towards understanding genetic relationships between certain other mutations.

North American Mutations

Bronze Fallow (a^{bz})

This mutation, which is generally known simply as 'Fallow', was established in the USA in 1971 (Alderton 1989). It is now widespread throughout Europe and the USA. Inherited as an autosomal recessive gene, Inte Onsman has reported breeding experiments in Europe which showed that the mutation is an allele of the rare NSL Lutino. Fallow mutations that are proven to be alleles of the NSL ino (tyrosinase) locus are now designated Bronze Fallow to distinguish them from other Fallow mutations (Ashen, Dun and Pale Fallows).

Bronze Fallow Cockatiel cock.

As an allele of the tyrosinase locus, the Bronze Fallow results in an altered tyrosinase enzyme which is the starting point for melanogenesis (the process of producing melanin). It is known from studies on similar mutations in mammals, that the mutant tyrosinase is generally altered in the way that it is folded during production and this is what changes its action in initiating melanin production. Typically in mammals and passerines, incomplete albinism of this locus results in failed eumelanin production but retained levels of phaeomelanin production. Yet Cockatiels, like other parrots, are currently thought to be unable to produce phaeomelanin. This is refuted by Professor Kevin McGraw (2006) who believes that all plumage contains at least traces of both melanin types. It is hoped that future research might determine the truth, as the shade of colour apparent in the Bronze Fallow could easily be explained by a significant phaeomelanin component.

Bronze Fallow Cockatiel.

Bronze Fallow chicks hatch with a red eye which is retained throughout life. The melanin in the plumage is altered qualitatively, becoming brownish in colour and much lighter than wildtype. Perhaps the best way to imagine the colour is to envisage a shade of pale beige part way between Cinnamon and Lutino. The brown tone to its colour helps distinguish Bronze Fallow from some of the other autosomal recessive mutations in this species. Psittacin pigments are unaffected and retain their normal distribution. Since the melanin pigment is so greatly reduced, the degree of yellow suffusion in the plumage becomes far more noticeable than in the wildtype.

It can, however, be confused with colour mimics produced by various combinations of melanin-altering mutations. Cinnamon-ino is a common combination that might be confused with Bronze Fallow; however Cinnamon-ino should retain far less melanin pigment. In Australia we have Fallow birds which have been established independently to the Bronze Fallow from the USA. As a result it is currently uncertain whether the Fallow that we have in Australia is a Bronze Fallow or not. Also, with the broad range of melanin-altering mutations that we have in Australia, some of these in combination will inevitably appear similar to Fallow morphs as well. In particular, the Cinnamon Faded combination is a similar shade of light brown to a Bronze Fallow and has a reddish eye, although it is not as clear red as that of the American mutation.

Desirable Matings

Like all melanin-altering mutations it combines best with the psittacin-altering mutations (Whiteface, Paleface, Yellowface and Yellowcheek) and the Opaline pattern gene. Being a light shade of colour, contrast in Pied specimens is poor. And as discussed already, combination with other melanin-altering mutations is fraught with danger, both in the correct identification of combination colours and the damage done to the purity of the gene pool of rare mutations.

Dominant Yellowface (T)

The Dominant Yellowface mutation appeared in Florida, in the USA, around 1996. The original breeder did not recognise what they had because adult cocks change their cheek colour and as a result the mutation became disseminated throughout the gene pool in that area. This led to multiple breeders suddenly 'discovering' that they had birds of this mutation in their collections over a short period of time. American breeders usually refer to this mutation as 'Dominant Yellowcheek' rather than Yellowface. However, I have chosen to use the term Yellowface as this is more consistent with the naming of the same mutation in a number of other species, as well as helping to distinguish it from the sex-linked Yellowcheek Cockatiel mutation from Europe.

A comparison of four different face and cheekpatch colours in Cockatiels. From left to right: Dominant Yellowface, Paleface, Normal and sex-linked Yellowcheek.

In my book, **A Guide to Colour Mutations and Genetics in Parrots**, I refer to this class of mutation as Tangerine, a name coined by Jim Hayward for the gene responsible for Orangeface in Peach-face Lovebirds. This mutation also exists in Fischer's Lovebirds in Australia where it is known as Yellowface. And there are also examples of this mutation in some lorikeet species, such as the Yellow-phase Dusky Lory. In all instances, the mutation is autosomal co-dominant with the homozygous (double-factored) phenotype being the best expression of the mutation. Many lovebird breeders even refer to single-factored Orangeface Peach-face Lovebirds as visual splits.

However, the similarities between all of these mutations is there for all to see. Basically, the gene responsible blocks the production or inhibits the use of 'advanced' psittacofulvin pigments. The bird still produces normal amounts of yellow psittacin, but red psittacin and often orange psittacin is greatly reduced or replaced. At present, we do not know for certain how this gene works. In fact, scientists are only just beginning to study the psittacofulvin pigments and still cannot even tell us where they are produced, although it is believed to be in the skin of the parrot. Perhaps study of these dominant Tangerine mutations might one day help determine the details of this process.

In lovebird species, the action of the gene results in a change of colour to the face area and tail markings, with no change to the green body colour because yellow psittacin is unchanged. In the Cockatiel, the only 'advanced' psittacin pigment is the deep orange cheekpatch and this becomes a yellow-orange colour, deeper in the cock than the hen. In adult cocks, the cheekpatch becomes so dark that it could be confused for a wildtype, which is why the mutation took time to be discovered.

Considering the understanding that we have of this mutation in other species, the likelihood is that a double-factored Yellowface would have a purer yellow face and be more likely to retain such in the adult cock. However, at present, the American breeders are concentrating on producing high quality youngsters via regular outcrossing through Normal birds and therefore nearly all specimens are single-factored.

Desirable Matings

The dominant Yellowface mutation is quite valuable for producing new combination colours. It can easily combine with all the different melanin-altering mutations as well as the Opaline pattern gene to basically create a new 'series' of colours parallel to Normal, Whiteface, Paleface and sex-linked Yellowcheek. I would not, however, recommend combination with the other psittacin-altering genes. Whiteface would mask the presence of dominant Yellowface as it leaves no psittacin pigment for Yellowface to work upon. Paleface could in theory be combined, but would merely result in a weaker-coloured Yellowface, which would be both less attractive and also potentially confusing. Word from the USA is that this has already happened to some degree, with breeders unexpectedly producing Paleface birds from some of the dominant Yellowface lines. Finally, combination with the sex-linked Yellowcheek would merely confuse the genetics of the birds without being expected to change the appearance very much. So without question, it is essential that Cockatiel breeders are careful to maintain Yellowface strains clean of other psittacin-altering genes.

Above: Single factor Yellowface Grey Cockatiel showing deeper cheek colouration typical in mature specimens of this mutation. This co-dominant mutation is expected to be more yellow in the double-factored form.

Below: Young single factor Yellowface Cockatiels in different colour combinations.

'Emerald'

The history of the 'Emerald' Cockatiel is reported in the American National Cockatiel Society's *Journal* (1998). Margie Mason of Texas began working with the mutation during the 1980s. The breeders who first found the new mutation in their aviary were Norma and John Ludwig and they asked Margie to work with the birds. A Normal cock was mated to a Normal hen—said to be brother and sister. This pairing produced three babies, one of which was visual for the 'Emerald' mutation. The second clutch produced the same results. These birds were first displayed at the National Bird Cage Show in New Orleans in 1995.

This mutation has also been sometimes known as 'Olive' in the USA and recently there have been moves by one of the two American Cockatiel Societies to designate the colour as a Suffused mutation. Unfortunately there has been much confusion regarding the nature of the yellow pigment in this bird, with many breeders believing that this gene increased the yellow in the bird along with the obvious melanin reduction. This has led to the mutation now being called 'Suffused Yellow' by many breeders.

It must be realised that the visibility of yellow pigment in this colour morph is due entirely to the reduction in melanin pigment, simply allowing the yellow to show through. The mutation is purely and simply a melanin-altering mutation. The term 'Suffused' was chosen because it is the name of a heavy melanin-reducing dilution mutation. It was then mistakenly combined with the word 'yellow' to reflect what many see as the appearance. However, if the Suffused mutation were to be correctly applied, the correct designation would be Suffused Grey, meaning that only a small diffusion of grey colours remained in the bird (the yellow being unaffected).

Unfortunately, it does not appear to me that this mutation is indeed a Suffused gene. As a dilution gene, Suffused does not alter the colour of soft tissues, instead affecting only the transport of melanin from the follicle into the growing feather. Eyes, beak and feet all remain normally coloured. Yet every photograph that I have obtained of the American 'Emerald' clearly shows reduced melanin pigmentation of these areas (either pale flesh-coloured or lightly pigmented). In contrast I suggest that the reader compare the photographs of this morph to the similarly coloured *Australian Suffused ('Olive')* mutation on page 149.

US breeders report that the 'Emerald' generally fledges darker in colour and then lightens as it matures. This feature is similarly reported for the Australian Suffused mutation, although other features appear to differ. As mentioned already, the 'Emerald' has a pale beak, and pale legs and feet in virtually all photographs that are available on the internet. Some show dark nails and a few show a degree of melanin in the legs, but no photographs that I have seen show pigmentation as dark as that seen

'Emerald' Cockatiel cock. The pale beak, legs and feet distinguish the 'Emerald' from the true Suffused mutation from Australia.

'Emerald' Cockatiel cock.

in the Australian Suffused mutation. US breeders report that the legs darken as the bird matures and some have suggested that the pale legs are due to the birds being split for multiple other mutations. What is clear is that there appears to be a visual difference between the Australian and American birds which indicates that they are two different mutations and that the American birds do not fit the criteria for a 'Suffused' mutation.

Then to add further to the confusion, by calling the new colour 'Suffused Yellow', US breeders inadvertently bring their mutation into confusion with the Yellow Suffusion trait which is independently the real cause of the variable yellow pigmentation, not just of the 'Emerald', but of all Cockatiel colour morphs.

Assigning or changing the name of a new colour morph is always a very delicate procedure. It is very important that any change be accurate, while, on the other hand, the use of incorrect names can become ingrained and difficult to change. In our current state of knowledge about parrot genetics and pigmentation, there is much that we do know and equally more still to be explained. It is simply not possible to instantly discover or determine the identity of many new colours. At present the 'Emerald' falls into this category and while this name is not really correct, at least it is not a defined term being incorrectly applied to this mutation. So for now I will refrain from making any suggestions about its identity until much more is learnt about the gene creating this important colour morph.

As explained already, the primary action of the 'Emerald' mutation is to greatly reduce the melanin pigment throughout the plumage. There is a tendency for specimens to retain deeper melanin head and shoulder colour than in the rest of the plumage. Because melanin is greatly reduced, any yellow pigmentation in the plumage is easily visible, which results in some strains carrying the Yellow Suffusion trait being very yellow indeed. The combination of strong yellow and a little melanin results in some breeders visualising an olive-green colour, hence the names 'Emerald' and 'Olive'. However, it must be understood that these birds never produce any true structural green colouration (as all Cockatiels are true Greygreen to begin with) and the colour has nothing to do with the true Dark factor gene that produces Olive in green-based parrot species.

Desirable Matings

'Emerald' would combine well with the psittacin-altering mutations (Whiteface, Paleface, Yellowface and Yellowcheek) and the Opaline pattern gene. Combination with melanin-altering mutations would likely result in pale, washed-out colour morphs that would be difficult to identify. However, test matings between these colours could help to accurately identify the 'Emerald' mutation. Considering the loss of melanin from soft tissue structures, it may prove to be an allele of the NSL ino locus and as such should be test mated to Bronze Fallow. All offspring from any experimental matings should never be allowed into the general Cockatiel population as they could eventually destroy this and other mutations.

There have been suggestions that European breeders combining 'Emerald' and Dominant Edged (Dominant Silver) have produced a bird being described as 'Black-headed' by some. It appears that the darker head of the 'Emerald' is accentuated by the Dominant Edged gene, perhaps merely by making the rest of the plumage even lighter.

'Goldcheek'

The name 'Goldcheek' has been given to a possible new mutation in the USA. Although this colour looks similar to a dominant Yellowface, it behaves genetically like a Parblue mutation. In other words, it is autosomal recessive and believed to be an allele of the blue locus, placing it into the same family as Whiteface and Paleface.

Because of the potential confusion caused when Paleface and Yellowcheek are interbred, it will take some time for breeders to determine if indeed this is a new Parblue-type mutation or merely a confused combination. To confirm this, it must be outcrossed through Normal Cockatiels that are known to be free of Whiteface and Paleface genes.

'Goldcheek' Cockatiel hen. There are still more questions than answers with this new mutation.

This in itself is a difficult thing nowadays with so much combining of different colours. The vast majority of Normal birds will be split for one of these two other colours.

However, if possible, a true breeding strain of 'Goldcheek' must be established, which can then be test mated under controlled conditions against the other psittacin-altering mutations. If eventually proven to be a new Parblue mutation, the 'Goldcheek' would create yet another series that all the melanin-altering mutations could be combined with.

Against this hypothesis, US breeder Laurie Bethea has conducted test matings between 'Goldcheek' and Whiteface and produced 24 Normal chicks without any Whiteface or 'Goldcheek' offspring. This result indicates that the 'Goldcheek' is not a Parblue mutation, but a mutation of a new locus that has not been identified in other parrots. Clearly more research is required and the importance of using pure strains of different colours for test matings and the development of new mutations is reinforced.

Australian Mutations

Faded ('West Coast Silver') (fd)

In the first edition of this book Cross and Andersen reported that this mutation appeared in Western Australia around 1982. The first breeder is not recorded, although Peggy Cross herself had a large role in its subsequent development, referring to it simply as 'Silver'. Sindel and Lynn (1989) do not record this mutation in their book. Since then other unrelated 'Silver' colours have appeared in Australia and Cockatiel breeders here now use the name 'West Coast Silver'. The mutation that creates this colour is correctly called Faded, although this name is largely unknown to the majority of Australian Cockatiel breeders.

The Faded mutation was originally discovered and defined as a colour morph by Budgerigar fanciers. It has now been identified in a number of species including the Swift Parrot, the Fischer's Lovebird and the Princess Parrot, as well as the Cockatiel. It is an autosomal recessive gene and it has the smallest degree of melanin alteration of all the albinistic mutations. Chicks are born with a slightly plum-coloured eye which darkens towards normal after a few days. Beak, legs and feet are also slightly lighter

Faded Cockatiel cock. This colour morph is generally known as 'West Coast Silver' by Australian breeders.

in colour than normal. An alteration to eye colour from normal at some point in life is the defining trait for all albinistic-type mutations. These are mutations which act to alter melanin quality through alterations to a stage of melanogenesis (the process of producing melanin).

To put this into simpler terms, the Faded mutation slightly alters melanin quality rather than quantity, which is why brownish tones are visible rather than just a pure grey. This feature led some early breeders to think that they had discovered the long awaited Cinnamon. And when the mutation first became available to the general public in the early 1990s it was often sold incorrectly as 'Cinnamon'. It is also the reason why Stan Sindel refers to this mutation in his classification system as 'Recessive Cinnamon'. However the gene causing this mutation is distinct from the Cinnamon gene and the phenotype is clearly different, being much closer to grey than to brown.

The actual mode of action for the Faded gene is currently unknown. Some authorities have suggested that it may be coding for an enzyme known as tyrosinase related protein-2 (TRP-2). However, mutations of that locus would be expected to produce a slate-grey colour phenotype without any brown tone and therefore seems unlikely.

Cross and Andersen (1994) described their earliest breeding results with this mutation: 'Photographs of the first nest that I bred that produced a 'silver' bird show a Lutino chick, a Normal (Grey) chick and a

Faded Cockatiel hen.

Faded Cockatiel cock.

Faded Cockatiel hen.

brown chick. The brown tones of the third chick are very obvious but the bird matured to be quite pale grey, especially on its chest.'

They also described a small degree of sexual dimorphism in the chicks: 'At fledging, cocks are usually lighter in colour than hens and will gain a few shades of darker grey on their backs when mature. The fledging colour is a guide to early sexing of this variety.'

Desirable Matings

Like all melanin-altering mutations, the Faded gene combines well with psittacin-altering mutations (Whiteface, Paleface, Yellowface and Yellowcheek) and the Opaline pattern gene. Selecting for Yellow Suffusion can detract from the overall colour as the human eye can misinterpret the juxtaposition of yellow and grey as a brownish colour. The true tone of melanin colour is best observed in the Whiteface Faded combination where all psittacin pigment is removed and no distortion of colour can occur.

Combinations with melanin-altering mutations should be made prudently. Although test matings have a scientific value, random mixing of the different recessive melanin-altering mutations will contaminate gene pools and lead to large numbers of unidentifiable colours. The general result of combining multiple genes from the same class is to move the colour produced towards lutino in steps. As the colour becomes lighter, accurate identification becomes more difficult.

The Faded Cinnamon combination has been produced and phenotypically it is difficult to distinguish from a Fallow colour morph. In the early days of the development of both the Cinnamon and Faded mutations in Western Australia, the two mutations were confused by some breeders despite their distinctive differences in appearance if studied closely. This has led to some Cinnamon strains being contaminated with the recessive Faded gene, which then appeared as Faded Cinnamon combinations in later generations of breeding. As a result, many Cockatiel breeders mistakenly thought that they had bred the Fallow mutation. This explains the appearance of 'Fallows' that Cross and Andersen reported in the first edition of this book being bred from Australian strains of Cinnamon Cockatiels. Cross and Andersen have since performed the necessary test matings to confirm this point.

The reason that this problem has appeared in Australia is that Faded and Cinnamon both appeared within a short time of each other in the same state of Australia and bird dealers often sell new colours by the wrong name. Breeders unfamiliar with the new colours generally have to accept the name that the dealer called the bird and this has led to many instances of incompatible mutations being interbred in many species of birds.

Platinum Cockatiel hen.

Platinum (Z ino^{pl})

The Platinum mutation is currently unique to Australia and appeared in the late 1980s. The mutation was developed in Brisbane, Queensland where large numbers were produced by a number of different breeders from the western suburbs. Many of the early specimens were Platinum Pearl combinations and the increased yellow pigmentation produced by the Opaline gene would have complicated interpretation of the colour of this mutation. Breeders in Queensland referred to the birds as 'Lacewing Cinnamon', keeping in mind that the true Cinnamon from Western Australia was unknown in the east of the country at that point in time. By the early 1990s when it was becoming available to the general public, Queensland Cockatiel breeders realised that it was a unique mutation and gave it the name Platinum.

The choice of name might cause some confusion for UK breeders as it was popular there to refer to the Whiteface Dominant Edged (Dominant Silver) combination by the name Platinum. It is therefore important for international breeders to recognise that the Australian Platinum is a unique primary mutation and not a combination of other colours.

The Platinum is a sex-linked recessive Parino mutation. This means that it is an allele of the sex-linked ino locus and it behaves co-dominantly when paired with the ino allele. It is a direct analogy to the Blue and Parblue mutations. As such, the Platinum in its action is a partial ino gene, producing a melanin reduction of approximately 50% from wildtype resulting in an even shade of light smoky grey as well as reduced melanin in the beak, feet and nails which are pale beige. When chicks hatch they have a plum-coloured eye which darkens towards normal by about one week of age. Psittacin pigments are unchanged by this melanin-altering gene. However, since the melanin is strongly reduced, the degree of yellow pigmentation in the plumage is more easily visible in the Platinum. As stated previously, this does not imply that Platinum is increasing the yellow pigments in the plumage in any way.

Platinum Cockatiel cock.

Similar mutations exist in other parrot species. In fact sex-linked Parino mutations now comprise a large group, having been found in numerous species including Budgerigars, Indian Ringnecked Parrots, Peach-face Lovebirds, Scarlet-chested Parrots and Red-rumped Parrots. The latter species actually has two different examples. European standards now refer to sex-linked Parino mutations as Pallid, although it is not certain that all examples in different species reflect the same gene. There is variation in the degree of melanin reduction in different species and some examples do not begin life with a plum-coloured eye. Therefore, it is likely that eventually we will discover many new Parino mutations that will form a large multiple allelic series with Lutino. As such, it is appropriate to retain the name Platinum as the internationally recognised name for this mutation in Cockatiels.

As an allele of the sex-linked ino locus, when Platinum is mated to Lutino, all sons produced will receive one platinum gene, one lutino gene and no wildtype genes. Hence they fall midway between Platinum and Lutino in plumage colour and the chick hatches with a ruby rather than a plum eye colour. The Australian National Cockatiel Society refers to this blended phenotype as 'Platino', although the generally recognised international term would be PlatinumIno or LutinoPlatinum, a simple blending of the two names without contraction. The blending indicates to the reader that the two different mutations are alleles and that the product is not a standard combination (which would have two genes for each mutation, not just one gene of each).

With respect to these mutations, Cockatiel cocks can carry the following genetic make-ups because they have two Z chromosomes:

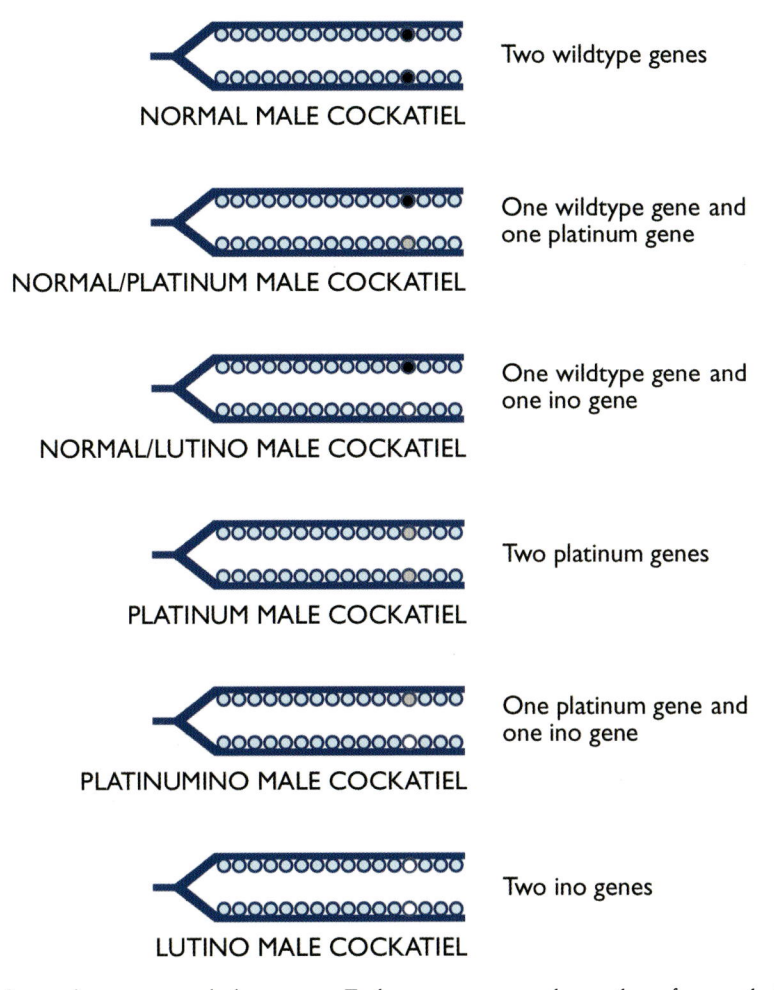

NORMAL MALE COCKATIEL — Two wildtype genes

NORMAL/PLATINUM MALE COCKATIEL — One wildtype gene and one platinum gene

NORMAL/LUTINO MALE COCKATIEL — One wildtype gene and one ino gene

PLATINUM MALE COCKATIEL — Two platinum genes

PLATINUMINO MALE COCKATIEL — One platinum gene and one ino gene

LUTINO MALE COCKATIEL — Two ino genes

Cockatiel hens, however, only have one Z chromosome and can therefore only have the following genetic make-ups:

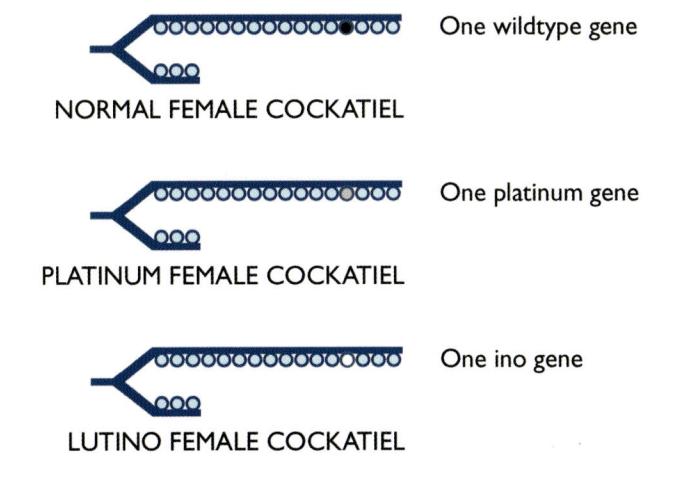

NORMAL FEMALE COCKATIEL — One wildtype gene

PLATINUM FEMALE COCKATIEL — One platinum gene

LUTINO FEMALE COCKATIEL — One ino gene

Platinum Cockatiel cock.

Platinum Cockatiel hen.

It can be seen from these genetic make-ups, that hens will always show which gene they are carrying and also that it is impossible to produce a PlatinumIno hen as she has no room on her single Z chromosome to carry two different alleles.

Mike Anderson, a Brisbane aviculturist who specialised in Cockatiels and was responsible for proving the allelic relationship between Platinum and Lutino, recalls that in the early days he was shown a bird which was claimed to be the 'first' Platinum. He identified this bird as a 'Platino', which raises a very interesting point. It is generally considered that sex-linked mutations will always become visual in hens first. However, in this case, the special relationship with Lutino would make a cock equally possible as a hen to be the first 'Platinum'. It also increases the likelihood that the Platinum mutation arose as a back mutation of the lutino allele, which is where a lutino allele mutates a second time to become a platinum allele.

Desirable Matings

Platinum and PlatinumIno both combine well with the psittacin-altering mutations (Whiteface, Paleface, Yellowface and Yellowcheek) and the Opaline pattern gene. There is enough retained melanin in the Platinum to produce visibly distinctive Pied combinations although the contrast is not as good as with darker mutations. However, this is not the case with the PlatinumIno. Combinations with other melanin-altering mutations are generally ill-advised and have largely been unexplored. It would be possible to produce a Cinnamon Platinum, which would be somewhat similar to a Bronze Fallow in appearance but without the bright red eye. Test mating of Platinum against new sex-linked recessive, melanin-altering mutations is advised to help determine their identity; however the progeny should be prevented from entering the general Cockatiel gene pool.

Confusion between Platinum and Cinnamon occurred when the two mutations first became available to the general public and this led to a degree of mixed matings with these two colours. Fortunately, due to the tight linkage of their loci, very few if any Cinnamon Platinum combinations were produced and neither gene pool suffered in their development. However, even today, I have not seen a photograph to document a true Cinnamon Platinum combination.

Dilute ('Pastel Silver') (dil^{gw})

The Dilute mutation appeared in a colony aviary at Dural, north-west of Sydney, New South Wales during the 1980s and was originally referred to as 'East Coast Silver', a name that was also sometimes applied to the Platinum mutation around this time. It is now generally known to Cockatiel breeders as 'Pastel Silver' although there is no evidence to suggest that it is a Pastel mutation. It is inherited as an autosomal recessive gene.

Greg Paull of Ipswich, Queensland purchased two original coloured birds and a 'split' in 1987 during a visit to a Sydney breeder whose name has not been recorded. Greg spent a number of years doing the groundwork to establish this mutation (Paull 1992) before a change of situation caused him to sell all of his stock to various breeders in Queensland in the early 1990s. At that time numbers were still small and the colour morph was virtually unknown outside of the South-East Queensland area. A few years later Mike Anderson of Brisbane, Queensland purchased a number of birds from Des Kitching who had been trying to develop the new mutation. The origin of Des's birds is not recorded. Mike set about outcrossing and improving the vitality of the mutation. Largely through his efforts, the Dilute mutation became available to the general public in the late 1990s.

Phenotypically, the mutation reduces melanin throughout the plumage to roughly 50% of Normal in an even pattern. Beak, eyes, legs and feet retain normal melanin content, indicating a true Dilute mutation. The degree of melanin reduction suggests a genetic parallel to the Greywing mutation in Budgerigars. However, to date, the mutation has not been studied microscopically to confirm this. And, of course, the name 'Greywing' could apply to nearly all Cockatiel colours and is therefore not suitable in this species. Hence I have chosen to retain Dilute as the name for this mutation in Cockatiels.

Dilute Cockatiel cock. P ODEKERKEN

Dilute Cockatiel hen. D ANDERSEN

Dilute Cockatiel. The true Silver Cockatiel is an Australian mutation generally known as 'Pastel Silver'. G ROMAN

True Dilute mutations act by interfering in the transfer of melanin from the melanocytes situated in the feather follicle into the new growing feather. Normal melanin is still produced in the skin and other soft tissues; it just cannot be used in normal quantities within feathers. They are therefore referred to as quantitative mutations instead of qualitative like the albinistic mutations.

The wildtype dilute allele can be mutated in a number of ways, creating an allelic series. This is known from Budgerigar breeding, where Greywing, Clearwing and Dilute (Suffused in Europe) form just such an allelic series. Different alleles of the gene differ in the quantity of melanin transfer and sometimes by the zones of the feather that the pigment is transferred to. Clearwing blocks transfer of melanin to cortical regions of the feather but allows normal melanisation of the medulla. This results in normal body colour but reduction of the foreground colour which produces the black mantle and wing markings of the Budgerigar.

Since the Cockatiel does not have a clear distinction between body colour and foreground pigmentation, it is difficult to predict how a Clearwing mutation would look in this species. What is certain is that it would not have clear wings like a Budgerigar, as the Cockatiel carries grey body colour throughout its wings, which the Budgerigar does not.

The name Pastel indicates an allele of the NSL ino locus that evenly reduces melanin colouration in the plumage and also reduces the melanin in the soft tissues such as beak, eyes and legs to some degree. To prove a Pastel mutation, apart from fitting the appropriate description, the gene must be test mated against an NSL ino gene or a Bronze Fallow gene and shown to be allelic. To date, none of the 'Silver' mutations in Cockatiels have been shown to be true Pastel genes. Breeders should therefore reserve this name for the time in the future when the true Pastel mutation is identified. Unfortunately, I expect that breeders will be reluctant to accept any name change for this colour morph.

The Dilute mutation produces the purest, mid-range silver colour possible in the Cockatiel. It is a colour that is desired by European and American breeders, who try to achieve similar results with their Dominant Edged mutation by selecting against the edging effect. The existence of the Dilute mutation in Australia reinforces the importance of breeding colours for their characteristic effects and not trying to select them towards our own desires. Otherwise we will end up with many different mutations all looking very much the same. Cockatiel breeders have been given the perfect chance to savour many more melanin-altering mutations than breeders of any other species can. As such they should accept the challenge to appreciate their unique differences by breeding to standards that reflect these qualities.

Desirable Matings

As discussed many times already in this book, combinations with other melanin-altering mutations should be done carefully to avoid contamination of gene pools and loss of distinctive mutations. Test mating to other mutations is a useful scientific goal, but should be done in a controlled way. It has been shown already that the Edged Dilute is not allelic with the Dilute mutation. The only other logical test mating of worth is to test the Australian 'Olive' mutation to confirm that it is a Suffused mutation.

Because the Dilute mutation retains significant melanin, it would produce a detectable Dilute Cinnamon combination although it may appear to be merely a 'poorly'-coloured Cinnamon specimen. Dilute Pied does make an attractive combination and even more so when combined into a Whiteface Dilute Pied. Dilute combines best with any of the psittacin-altering mutations (Whiteface, Paleface, Yellowface and Yellowcheek) and the Opaline pattern gene.

Australian Fallow (f)

Sindel and Lynn (1989) report that this mutation first appeared in the 1960s, with a number of breeders being involved in its development including Fred Lewitska, Joe Mattinson and Ron Rodda. This Australian Fallow mutation therefore predates the appearance of the US Fallow in 1971. It is therefore possible that the Bronze Fallow (USA) may have descended from Australian birds or may have arisen independently.

As there is currently no legal movement of these birds into or out of Australia, we cannot compare the two colour morphs to determine if they represent one or two different mutations. And since we do not have an NSL Lutino in Australia, our bird cannot be test mated to determine its locus. To add to the uncertainty, many of the birds currently being bred in Australia and identified as 'Fallow' may in fact be Faded Cinnamon combinations.

Because the Australian Fallow has a unique origin, unrelated to the American birds, it would be a shame if it were lost through replacement by a combination colour mimic. I would strongly recommend all Australian Cockatiel breeders with Fallow birds to test mate them to identify any Faded Cinnamon birds and separate them from their Fallow gene pool. Fallow cocks should be outcrossed to Normal Grey, which should produce only Normal Grey split Fallow young. If Cinnamon daughters are produced, the bird is a Faded Cinnamon. Fallow hens should be test mated to Cinnamon cocks to produce only Grey split Cinnamon split Fallow son. If Cinnamon sons are produced, then once again the bird must be a Faded Cinnamon.

Considering that the true Australian Fallow was reported to be difficult to breed, it may be that it has already been lost and replaced by the Faded Cinnamon combination which is as easy to breed as the two common mutations that it is produced from.

Like all Fallow mutations, chicks hatch with a bright red eye which is retained throughout life. The beak, legs and feet have virtually no pigment and appear an extremely pale tone of grey-beige, while the toenails are pale beige. Cross and Andersen (1994) described the Australian Fallow 'as existing in two colour phases'. One is described as being similar to a Cinnamon-ino with

Australian 'Silver' Fallow hens. Is this the real Australian Fallow?

Australian Fallow Cockatiel hen.

Australian Fallow Cockatiel cock.

Are these birds Bronze Fallow or merely a combination of other mutations? Clearly they do not have the bright red eye of the American Bronze Fallow.

both sexes alike, which is also similar to the Bronze Fallow from the USA. However, I suspect that these might be the Faded Cinnamon combination birds. Most of the photographs in the original edition of this book were of these birds.

The second 'colour phase' is described as 'a delicately coloured mutation with tones of pale grey suffused with hints of brown, mainly on the backs of the birds'. This other colour morph is also sexually dimorphic. Cross and Andersen (1994) describe it as follows:

'The hen of the silvery-brown phase has to be described as a two-tone bird. Her face is dilute yellow with no grey masking effect. The same yellow tones continue down her chest but are more dilute. Viewed from the back she is an entirely different colour. Her back is mottled or shaded in grey tones with a hint of brown—lighter than the 'Pastel' Silver birds. Silvery-brown phase cocks have more grey tones on their backs and their fronts are pale grey with less yellow than the hens.'

The accompanying photographs of this colour morph indicates that it is a distinctly different type of Fallow from either of the overseas Fallow mutations. I suspect it may be the true original Australian Fallow. Unfortunately, all the 'Fallow' Cockatiels that I have seen in recent years resemble the first type described and not this more silvery-brown type of Fallow. As stated already, I fear that the true Australian Fallow may have been lost. Anyone with specimens fitting this silvery-brown description should endeavour to reproduce them and keep them pure.

Desirable Matings

The Australian Fallow should only be combined with the psittacin-altering mutations (Whiteface, Paleface, Yellowface and Yellowcheek) and the Opaline pattern gene. As previously stated, rare genes like Australian Fallow should not be mated to other melanin-altering mutations due the risk of polluting gene pools with recessive genes that might eventually result in the loss of the mutation.

Edged Dilute ('Silver Spangle') (ed)

The autosomal recessive Edged Dilute mutation appeared in Western Australia in the early 1980s (Cross & Andersen 1994). Records show that the first specimen was bred from a Normal (Grey) x Lutino mating. Despite being available for such a long period of time it was slow to be fully established due to excessive inbreeding in the early years and is only just now becoming more widely available. Hank Jonker has been heavily involved in the development of this colour morph.

Some of the early difficulty in breeding was partly due to confusion between this mutation and the Dominant Edged mutation available elsewhere in the world. Due to similarities in appearance, many Australian Cockatiel breeders made the incorrect assumption that this new mutation was one and the same with the European morph. It was therefore commonly referred to as 'Dominant Silver' for many years and even today some breeders still believe it is a dominant trait, even though Peggy Cross conclusively proved its recessive inheritance in the early 1990s.

The results of the test matings were recorded in the first edition of this book. Peggy Cross records: 'I outcrossed all my 'Silver Spangle' to Normal (Grey) and the subsequent breeding records of 1991 show that 45 offspring were produced. None showed the characteristic silvery, spangled, shaded markings.' The fact that no mutation young were produced confirms this colour as autosomal recessive and negates any suggestion of single and double factors being involved.

She continues with 'the next generation required the mating of 'Silver Spangle' cocks with hens bred from one of the other 'Silver Spangle' cocks. No daughters were bred back to their fathers. Two pairs bred and both produced 'Silver Spangle' birds. Finally we produced a 'Silver Spangle' hen.'

'The second pair produced 14 youngsters. At least seven of the chicks were visibly marked. Three birds displayed only slight markings. Four of the young birds showed a clear visible expression of the desired silvery, shaded effect. Two of these, both cocks, were extremely light in colour with the shaded silvery colour interference intruding through and obscuring the white wing bar. Two of the youngsters were hens! These young hens were not as spangled or as light in colour as the cocks.'

The Edged Dilute mutation acts to reduce melanin pigmentation to a large degree, but more significantly, it does not do this evenly. As a result a darker edge of melanin is retained on most feathers and this is most noticeable in the larger flight feathers. The amount of edging can be altered to some degree and even between barbs on the one feather vane. I believe that breeders should promote their traits as the main feature of this mutation and select against specimens with poor edging.

The first Edged Dilute mutation was established in the Peach-face Lovebird where it was originally known as 'American Yellow'. Despite being well established in this species, the mutation has not appeared in too many other parrots. Therefore, its appearance in Cockatiels is an important occurrence and all endeavours should be used to maintain it.

Edged Dilute Cockatiel cock. This colour morph is generally known as 'Silver Spangle' by Australian breeders. The Edged Dilute is an uncommon mutation in aviculture.

As a rare type of colour mutation across aviculture, little is understood regarding the gene action of this mutation. Although it is generally considered a Dilute mutation, some breeders have reported observing eye colour changes shortly after hatch in lovebirds, which suggests an albinistic-type mutation.

Cross and Andersen (1994) describe the chicks as follows: 'The eyes are dark brown at hatching and as adult birds. The feet and toenails are dark grey like those of Normal birds.' This description fits the Edged Dilute classification.

Desirable Matings

Edged Dilute combines well with psittacin-altering mutations (Whiteface, Paleface, Yellowface and Yellowcheek). Combination with the Opaline pattern gene is perhaps less desirable than for other melanin-altering mutations due to the conflict between the pearl and edging patterns. So far breeders have chosen to keep Opaline away from Edged Dilute Cockatiels and this seems wise.

Like other rare melanin-altering genes, it is important to keep its gene pool free of contamination by other similar colours. Unfortunately problems with this have already occurred, as breeders have reported producing other colours from outcrosses and test matings. Indeed, one of the causes of early confusion with this mutation was the appearance of lighter and darker specimens of Edged Dilute. Some breeders thought that the lighter 'phase' must represent 'double-factored' specimens. However it is now believed that the lighter 'phase' birds may be Faded Edged Dilute combination specimens.

The Dilute mutation has also recently been test mated to Edged Dilute by Brian Higginbotham of Victoria, producing Normal offspring in the first generation, thereby confirming that they are not allelic. The subsequent combination of the two mutations has produced a bird with greatly washed-out wings, virtually eliminating the edging characteristic of the Edged Dilute. It is to be expected that similar effects would occur with combinations of other melanin-altering mutations and breeders should be concerned that these experiments are kept isolated from the general Cockatiel population.

Edged Dilute Cockatiel cock.

Edged Dilute Cockatiel hen.

Suffused Grey Cockatiel pair, hen on right.

Suffused ('Olive') (dil)

The autosomal recessive mutation known in Australia as 'Olive' appeared towards the end of the 1990s and was established by Mike Anderson of Brisbane from two split cocks obtained from the original breeder. As the gene pool was so small—in effect starting with only 'half a mutation—Mike Anderson did an excellent job with careful outcrossing to establish this new Australian mutation.

The name 'Olive' has been used by Australian breeders due to the superficial resemblance of our birds to the American mutation known variously as 'Emerald', 'Olive' and now 'Suffused' as well. It is unfortunate that US breeders have adopted this latter term, as it appears to apply to the Australian but not the American mutation.

All of the melanin-altering mutations reduce melanin pigments in some manner. As the degree of reduction becomes greater, it is natural for all types to look more and more similar. This is what has caused confusion between the Australian 'Olive' and the North American 'Emerald'. Added to this, breeders have mistakenly focussed upon the visible yellow pigments in the bird, which are merely becoming more visible due to the loss of melanin and have nothing significant to do with either mutation.

In basic terms, both mutations heavily reduce melanin more than any other mutations so far known in Cockatiels, yet retain a silver-grey colour and dark eyes as adults. However, the true distinction between the two mutations can be seen in the colour of the beak and the legs. The Australian mutation is a true Suffused mutation as these structures are as dark as a wildtype Cockatiel. In contrast, the US 'Emerald' has a pale beak, and pale legs and feet which indicate a form of albinism. As discussed in the 'Emerald' section, I am not currently able to assign a definitive name to this mutation, but I do not believe that the term 'Suffused' should be used.

Australian Cockatiel breeders have been calling their colour by the name 'Olive', which is a mutation that cannot occur in Cockatiels due to their lack of structural colouration. However, breeders were wisely reluctant to change the name until the correct identity was found. Having studied the mutation in more detail in preparation for this book, I am convinced that it fits the definition of a Suffused mutation. It now requires further investigation to see if it fits all criteria.

Suffused is a mutation classification defined by European breeders. It is the lightest form of Dilute mutation, characteristically retaining only 10–15% of normal melanin deposits within the feathers. Being a Dilute gene, it has full normal melanisation of all other tissue including eyes, beak, legs and feet. This is due to the action of Dilute genes

Left: Suffused Grey Cockatiel cock—one of the early specimens bred by Mike Anderson. This colour is still widely known by the colloquial name 'Olive'.

Suffused Grey Cockatiel cock.

which control transfer of melanin from the melanocyte situated within the feather follicle into the new growing feather. Melanin is still produced normally, it just cannot move into the feather in normal quantities.

The mutation was previously known to exist in Budgerigars, Peach-face Lovebirds and Indian Ringnecked Parrots. To confirm the identity, ideally feathers need to be studied microscopically. I would also suggest test mating with the Australian Dilute mutation ('Pastel Silver') because of the possibility of an allelic series. If this was found to be the situation, then the first cross would produce an intermediate colour rather than Normal birds.

The Suffused Cockatiel is characteristically lighter on the wings and flights than on the back. Unusually for the Cockatiel, the front of the bird is darker than the back of the bird. Hens appear darker than cocks and cocks actually lighten in colour when they mature—which is the reverse of what is seen with all other mutations in this species.

Desirable Matings

Like all melanin-altering mutations, the Suffused mutation combines best with genes from other classes—the psittacin-altering mutations (Whiteface, Paleface, Yellowface and Yellowcheek) and the Opaline pattern gene. Apart from strict scientific test matings, it would be inappropriate to mix other melanin-altering genes into the Suffused gene pool. The result of such combinations would be colours approaching the clear phenotype, which would lack the distinctive features of either of the parent mutations. The Suffused Cinnamon combination colour already produced demonstrates this point quite well.

Pewter Cockatiel hen.

P. ODEKERKEN

J WATTS

Pewter (Z^{pw})

The Pewter Cockatiel is one of the latest mutations to be developed in Australia, appearing in 1998. It is a sex-linked recessive mutation that has proven to be independent of both the sex-linked ino and the cinnamon loci, indicating that it is indeed a new mutation currently unknown in other parrot species. As such, it has been given a distinctive name that does not conflict with any other recognised mutation and also happens to somewhat reflect the appearance of the bird.

The Pewter mutation alters the melanin pigment qualitatively, changing the shade of grey in a way difficult to explain. It has a slight brown tone but is still distinctively grey, slightly lighter in colour than normal but darker than that of any other established mutation. Because of the brownish tone in certain light, some breeders may confuse it for a Cinnamon. However, there are obvious differences apart from the fact that the plumage is not pure brown as it must be for a Cinnamon. The beak, feet and nails are also coloured differently from those of a Cinnamon Cockatiel.

The original bird was a cock discovered by Lawrence Jackson amongst a collection of Cinnamon birds offered for sale. After establishing the mutation, Lawrence Jackson performed test matings against both Cinnamon and Platinum, producing Normal sons in both instances to confirm that Pewter is indeed a new mutation. Some other Australian breeders have speculated that the bird

Whiteface Pewter (left) and Whiteface Cinnamon Cockatiel chicks. The removal of yellow by the blue (Whiteface) gene clarifies the true colour of the Pewter and the Cinnamon. This photograph clearly shows the difference in colour between these two sex-linked mutations.

Above: Pewter (left) and Cinnamon Cockatiels. Note the difference in the colour of the feet.
Right: Pewter Cockatiel hen.
Below: Pewter Cockatiel cock.

may have been a combination of Cinnamon and another mutation; however this has been completely disproved by the test matings performed to date.

As a unique mutation, we cannot speculate too much upon how the Pewter acts. The only other unique sex-linked mutation known in other parrot species is the Slate mutation in Budgerigars and that mutation acts to alter feather structure which affects structural colour thereby creating a greyish looking bird. It seems unlikely that this effect could be attributed to the Pewter but only time will tell.

Desirable Matings

The Pewter offers Australian Cockatiel breeders a new avenue for exploration, with only the Whiteface Pewter combination having been produced to date. Potentially it can be combined well with all the psittacin-altering mutations and the Opaline pattern gene. And being a dark morph, it will produce distinctive Pied combinations. Combinations with other melanin-altering mutations are probably only of value for investigative genetic testing, as they will generally be difficult to identify correctly. As such I recommend keeping Pewter strains free from other melanin-altering mutations, particularly the autosomal recessive genes.

Australian 'Yellowface'

A new mutation has appeared in Australia which produces a yellow cheekpatch instead of the normal orange cheekpatch. In appearance, it appears similar to both the dominant Yellowface and the recessive 'Goldcheek' mutations from the USA. The new colour was discovered in 2001 in an aviary containing Whiteface Cockatiels belonging to a breeder from Footscray, Melbourne and is currently being established by another Victorian breeder. The first known specimen was a coloured hen.

A significant number of offspring have been produced from the first generation which, at first thought, would indicate a dominant inheritance. However the mate used for the first outcross was split for Whiteface and a number of Whiteface offspring have also been produced. This means that the results could also indicate a recessive Parblue-type mutation which would be allelic with Whiteface. Results from the next generation of matings will hopefully confirm more about the inheritance and identity of this new Australian mutation.

Australian 'Yellowface' Grey Cockatiel hen.

A New Australian Mutation?

Among the 'Yellowface' offspring produced to date is a Lutino 'new colour' combination hen which shows virtually no yellow pigmentation in the general body plumage. This result suggests that the new colour is reducing yellow pigment throughout the plumage but not the face area and the effect is quite distinctive compared to the Paleface Lutino combination. This might indicate a Turquoise-type Parblue mutation, since Turquoise mutations reduce psittacin pigments unevenly.

Like all mutations, patience is required until the true identity is confirmed. Once established and identified, this mutation will add a new series to Australian colours in Cockatiels, being able to be combined with all the melanin-altering mutations as well as the Opaline pattern gene. It is advisable to keep psittacin-altering genes like Whiteface and Paleface away from this new mutation as they will potentially mask and confuse the appearance of the new colour. This makes it doubly unfortunate that the first outcross was carrying Whiteface, highlighting the importance for mutation breeders to maintain pure gene pools of the different mutations, rather than the current practice in some areas of aviculture to be the first to combine as many different genes as possible.

Australian 'Yellowface' Lutino hen.

COMBINATION COLOURS

Melanin Combinations

As discussed in the previous section, breeders should be cautious when considering the combination of different melanin-altering mutations. As we combine multiple genes that reduce the melanin content of feathers, the appearance gradually approaches that of a Lutino. Many combinations can mimic other colours, either primary mutations or combinations. The accompanying photographs illustrate some of the different combinations that have been produced to date.

Whiteface Dilute Edged Dilute Cockatiel cock.

Above: Dilute Edged Dilute Cockatiel cock. This combination loses some of the distinctive features of the Edged Dilute mutation.

Cinnamon (left) and Faded Cinnamon Cockatiels.

Left and right: This colour is known as 'Black-headed' in Europe. Many breeders believe that it may be a Dominant Edged 'Emerald' combination. The tendency for darker head colour, seen individually in these two mutations, appears to result in a unique retention of grey in the head of the combination.

Cinnamon Suffused Cockatiel hen. Many combinations lose their distinctive features as their colour gradually moves closer to the 'clear' phenotype.

PlatinumIno Cockatiel cock. Known also as 'Platino' or LutinoPlatinum by Australian breeders, this colour is halfway between Platinum and Lutino as it has one gene of each mutation. As such, it can only occur in cocks because only they have two Z chromosomes.

Above: The Faded Cinnamon Cockatiel could be confused with a Fallow.

Right: Cinnamon-ino Cockatiel cock. This is a subtle colour that nevertheless appeals to many breeders.

Whiteface Cinnamon-ino Cockatiel hen.

Pearl Combinations

The Opaline gene produces many of the most attractive colour combinations in Cockatiels. Opaline increases yellow psittacin pigment throughout large areas of the plumage and this can distort what the human eye sees, as well as how the mind interprets the colour of melanin pigments. In general terms, the addition of yellow pigment will complement how the mind interprets brown tones, but can make grey tones appear more brownish than they are. Some breeders will find these shades appealing while other people might prefer the Whiteface Pearl combinations due to their elimination of this effect. These latter combinations are illustrated in the section on Whiteface combinations.

Faded Pearl hen.

Faded Pearl hen.

Dilute Pearl Cockatiel hen.

Platinum Pearl Cockatiel pair.

Page 156

Cinnamon Pearl Cockatiel hen.

Lutino Pearl Cockatiel hen.

Bronze Fallow Pearl Cockatiel hen bred in Germany.

Cinnamon-ino Pearl Cockatiel hen.

Left and right: Single factor Dominant Edged Pearl Cockatiel hens.

Pied Combinations

Most people would consider Pied combinations to be most attractive when the contrast is greatest and hence the 'better' combinations involve Pied and the darker colours. However, other breeders love to explore the subtle colours of lighter Pied combinations. Like the Opaline gene, the Recessive Pied mutation has a tendency to increase the yellow pigments in the plumage and this can detract from the colour of some of the melanin-altering mutations.

Cinnamon Pied Cockatiel cock.

Dilute Pied Cockatiel cock.

Bronze Fallow Pied Cockatiel.

Faded Pied Cockatiel cock.

Edged Dilute Pied Cockatiel cock.

Pewter Pied Cockatiel.

Page 158

Grey Pearl Pied Cockatiel hen.

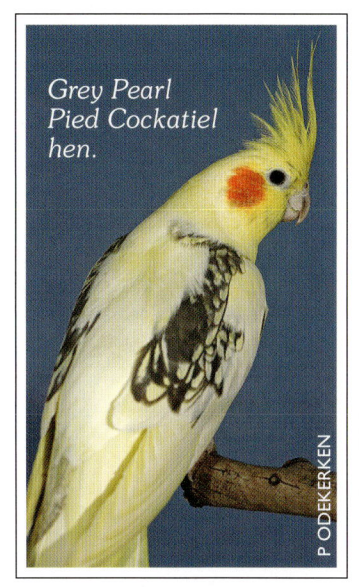

Pearl Pied Combinations

Pearl Pied was one of the very first combinations produced in Cockatiels. It is still popular amongst breeders to combine Pearl Pied with various other mutations. However, with advancements in Pied selection, the preferred Pied specimens have only small areas of remaining melanin and therefore only small areas to show the pearl pattern. The better the Pied, the fewer Pearl characteristics we get to see.

Cinnamon-ino Pearl Pied hen. A very subtle colour combination for the Pied enthusiast.

Above: Faded Cinnamon Pearl Pied Cockatiel. This combination can be incorrectly identified as a Fallow Pearl Pied.

Grey Pearl Pied Cockatiel cock. The anti-dimorphism effect of the Recessive Pied gene can result in some retention of pearls in cocks of this combination.

Lutino Pearl Pied Cockatiel hen.

Cinnamon Pearl Pied Cockatiel hen.

Faded Pearl Pied juvenile Cockatiel cock.

Whiteface Combinations

Whiteface Platinum Pearl hen (left) and Whiteface PlatinumIno cock.

Whiteface Cinnamon Cockatiel cock.

Whiteface Cinnamon Cockatiel hen.

Whiteface Platinum Cockatiel hen.

Whiteface Platinum Cockatiel cock.

Whiteface Faded Cockatiel cock.

Whiteface Australian Fallow Cockatiel cock.

Whiteface Bronze Fallow Cockatiel bred in Germany.

Whiteface Edged Dilute Cockatiel cock.

Whiteface Dilute Cockatiel cock.

Whiteface Combinations

Whiteface Pewter Cockatiel hen.

Whiteface Suffused Grey Cockatiel cock and hen (right).

Whiteface Suffused Grey Cockatiel cock.

Left: Whiteface Pewter Cockatiel cock.

Whiteface Cinnamon (left) and Whiteface Pewter Cockatiels. Note the feet colour.

Whiteface single factor Dominant Edged Cockatiel cock.

Whiteface 'Emerald' Pearl hen (left) and Whiteface 'Emerald' cock.

Whiteface single factor Dominant Edged Cockatiel hen.

Above: Whiteface double factor Dominant Edged Cockatiel hen. Right: Whiteface Lutino (Albino) Cockatiel. This combination is the only true Albino possible in the Cockatiel.

Whiteface Pearl Combinations

Whiteface Faded Pearl Cockatiel hen.

Whiteface Platinum Pearl Cockatiel hen.

Whiteface Dilute Grey Pearl Cockatiel hen.

Whiteface Grey Pearl Cockatiel hen.

Whiteface 'Emerald' Pearl Cockatiel hen.

Whiteface Cinnamon Pearl Cockatiel hen.

Whiteface Pied Combinations

Whiteface Grey Pied Cockatiel.

Whiteface Cinnamon Pied Cockatiel hen.

Left: Whiteface Ashen Fallow Pied Cockatiel hen.

Whiteface Platinum Pied Cockatiel hen.

Whiteface Cinnamon-ino Pied Cockatiel hen.

Page 165

Whiteface Pearl Pied Combinations

The blue gene produces an entire series of Whiteface colours in Cockatiels, basically doubling the colours that aviculturists can breed in this species. The human mind loves contrast and the removal of psittacin pigments enhances the contrast of colours in the Cockatiel. The removal of yellow pigments also aids our visual interpretation of melanin colours, making Whiteface combinations the best to compare between different mutations.

Whiteface Cinnamon Pearl Pied Cockatiel hen.

Whiteface Platinum Pearl Pied Cockatiel hen.

Left: Whiteface Grey Pearl Pied Cockatiel cock.

Whiteface Ashen Fallow Pearl Pied Cockatiel cock.

Whiteface Grey Pearl Pied Cockatiel hen.

Page 166

Paleface Combinations

The Paleface forms a third series of colours for Cockatiels, falling midway between Whiteface and Normal series birds in colour. The colours are more subtle than the other series, but nonetheless quite attractive. The differences are best appreciated when compared directly to Normal series or Whiteface series colours.

Paleface Cinnamon Cockatiel cock.

Paleface Platinum Cockatiel hen.

Paleface Lutino Cockatiel cock. This colour is sometimes referred to as 'Creamino'.

Paleface Dilute Grey Cockatiel cock.

Paleface Grey Pied Cockatiel.

Paleface Clear Pied Cockatiel cock. Selection has allowed breeders to develop this colour from the typical Pied mutation.

Paleface Platinum Pied Cockatiel cock.

Paleface Cinnamon Pied Cockatiel cock.

Paleface Platinum Pearl Pied Cockatiel hen.

Paleface single factor Dominant Edged Pied Cockatiel hen.

Paleface Grey Pearl Pied Cockatiel.

Paleface single factor Dominant Edged Cockatiel cock.

Paleface single factor Dominant Edged Cockatiel cock.

Paleface Grey Pearl Cockatiel hen.

Paleface Platinum Pearl Cockatiel hen.

Yellowface Combinations

The dominant Yellowface mutation forms the fourth series of colours for Cockatiels. The enhancement of yellow pigments by either Opaline or Pied can complement the yellow pigment of this mutation. As a new mutation, many of the possible combinations have yet to be explored.

Young Yellowface Grey Pied Cockatiel.

Above: Normal Grey (left) and Yellowface Cinnamon Cockatiel pair.

Yellowface Grey Pearl Cockatiel chick.

Yellowface Cinnamon Pied Cockatiel hen.

Yellowface Cinnamon Pearl Cockatiel hen.

Young Yellowface Grey Pearl Cockatiel.

Page 171

Yellowcheek Combinations

The sex-linked Yellowcheek mutation forms the fifth series of colours for Cockatiels. These can be difficult to distinguish from the Yellowface combinations unless the owner knows the genetic history of the birds. In general terms the sex-linked Yellowcheek produces a better yellow cheekpatch colour than that of the dominant Yellowface birds, which becomes more orange-yellow as they age. Until this mutation appears in Australia, many of the possible combinations cannot occur.

Yellowcheek Cinnamon Pearl Cockatiel hen.

Yellowcheek Grey Pied cock.

Yellowcheek Grey Pearl Cockatiel hen.

Below: Normal Cinnamon (left) and Yellowcheek Cinnamon Cockatiel hens.

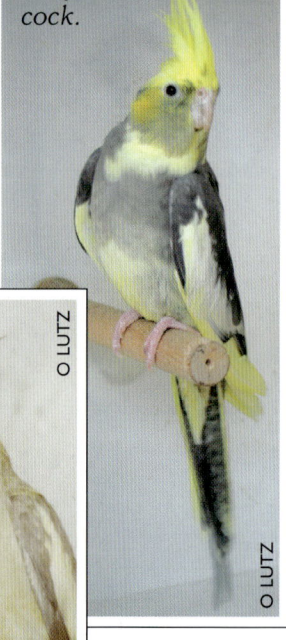

Yellowcheek Grey Pied cock.

Page 172

Yellowcheek Cinnamon Pearl Cockatiel hen.

Yellowcheek Cinnamon Pearl Pied Cockatiel hen.

Yellowcheek Grey Pearl Pied Cockatiel hen.

Yellowcheek Lutino Pearl Cockatiel hen.

Yellowcheek Cinnamon Cockatiel cock.

Yellowcheek single factor Dominant Edged Cockatiel.

COLOUR ODDITIES

'Halfsiders'

Like many other species of parrots in aviculture, the Cockatiel has produced specimens referred to as 'halfsiders'. A 'halfsider' is a bird that expresses different colour mutations in different regions of the plumage. The classic example is half one colour and half another colour, split down the centre of the bird. However many variations occur, including birds referred to sometimes as 'quartersiders' which have patches of disparate colour in different regions of their plumage.

In some species of parrots, the majority of 'halfsiders' appear to involve the chromosome carrying the blue locus, with specimens believed to be genetically Green split Blue. These birds will be half green series and half blue series, sometimes with the dark factor locus also involved as it is carried on the same chromosome. Currently the most credible theory is that one chromosome of the pair has been lost or damaged on the side expressing the recessive Blue gene.

Other 'halfsiders' involve expression of the Z chromosome and these specimens often show physical evidence of being gynandromorphs, ie the birds are half male and half female. The loss of function of one Z chromosome from the female side of the bird allows hidden sex-linked traits to express themselves.

This is my interpretation of what has happened to the bird in the photographs to the right, where a Whiteface Pied is showing pearl markings on only one side. The chromosomes carrying the Whiteface and Pied genes are unaffected. It is typical for 'halfsiders' to only show changes in genes from a single chromosome.

'Halfsiders' do not pass on their appearance genetically. In the case illustrated, if this bird bred it would reproduce as a Whiteface Pied split Pearl cock. However, if it is indeed a gynandromorph then it is most likely to be infertile.

Above and below: This 'halfsider' appears Whiteface Grey Pearl Pied on the left side of its body and Whiteface Grey Pied on the right side.

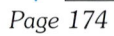

Left and below: This Australian-bred bird is a very unusual specimen. It has indications down the chest of a 'halfsider' effect. However, while the overall mottling pattern varies in degrees on each side of the wings and back, there is not the apparent expression of different genes on each side that are seen in typical 'halfsiders'. The pattern of mottling is reminiscent of one European Schimmel specimen. The broken pattern of pigment and white areas throughout the plumage might indicate a Mottled mutation. If so, Mottled is not a stable pattern and changes with each and every subsequent moult. Only time will tell if this anomaly will reproduce itself.

Schimmel

German breeder, Günter Wulf, has produced two birds that carry grey, brown and white in their plumage. He refers to these birds as 'Schimmel'—a German word which is also used to describe a dapple (or roan) horse, ie one with mottled markings of another colour on its coat. Canary breeders use the term as well and translate the word as 'mould'. These birds have a mottled plumage and it could be said that the mottling resembles mould growing on a white background. However, neither translation explains all the colour changes in these 'Schimmel' birds.

Above: The name 'Schimmel' is being used by the breeder of this bird. It might possibly be a form of Grey Pied split Cinnamon 'halfsider'.

One bird appears to be explainable as a Grey Pied split Cinnamon 'halfsider'. However, the second bird is more problematic with apparent blending of colours in some regions of the plumage. Perhaps the term 'mosaic' might more aptly describe these birds. Although these birds have bred and produced 'normal' colours in their offspring, so far the unusual 'Schimmel' colour has not been reproduced.

This 'Schimmel' has blended colours and might best be described as 'mosaic'.

Acquired Colours

Acquired colour changes are those that are not genetically controlled and are generally acquired after birth. They may appear in nestling plumage or may be attained later in life. These changes will sometimes disappear at different times or may be a permanent change due to the health status of the bird.

The most common acquired colour change in Cockatiels involves the spreading of orange colouration over the face area of breeding birds. The exact cause or nature of the change has not been explored scientifically but appears to involve orange psittacin pigments spreading from the cheekpatch. It would be interesting to note whether Yellowface or Yellowcheek mutations ever develop this trait. If they do not, then it would add support to the colour being produced by psittacin pigment; if they do, then the colouration must be caused by some other substance.

This Cockatiel shows orange smudging around the face and head, common for this species when breeding.

Page 175

Note the orange smudging around the face and head of these breeding Cockatiels.

Typically, this colour change is seen in adult Cockatiels while breeding and raising young. It seems to appear more commonly in those birds breeding for extended periods. It might possibly be hormonal or perhaps it occurs as a result of the bird being 'stressed by breeding'.

The other common acquired colour change seen in Cockatiels involves a general 'yellowing' of the plumage. The yellow colour is added throughout the plumage in a spreading manner, having no effect on the pre-existing pigmentation of the feathers. This syndrome has been linked by veterinarians to chronic liver disease and it is currently believed that bile pigments from the liver are responsible for the colouration. The colour can be added to mature feathers, not just growing feathers.

Often birds with this change are fed poor diets and suffer from fatty liver disease. They generally develop the colour slowly and often their owners do not recognise the change happening until it is very advanced and pointed out to them. Few birds ever recover fully from this condition and generally succumb to liver failure after an extended illness that might last several years. Beak changes and overgrown nails are also connected to this disease syndrome.

Above: Yellow pigment can be seen clearly in the chest area as it suffuses through the plumage of this bird suffering from hepatic lipidosis.

Left: Yellow pigment can be seen down the back of this Cinnamon Cockatiel hen suffering from hepatic lipidosis.

Right: This Cinnamon Cockatiel cock has extreme acquired yellow suffusion due to fatty liver syndrome. The badly overgrown and thickened beak is another common trait of this condition.

Page 176

Cockatiel Mutation Technical and Common Names

Gene Name	Colour Name	Incorrect and Former Names
ino	Lutino	'Albino', 'Yellow', 'Moonbeam'
cinnamon	Cinnamon	N/A
recessive pied	Pied	'Dominant Pied'
blue	Whiteface	N/A
aqua	Paleface	'Pastelface'
opaline	Pearl	'Lacewing'
dominant edged	Dominant Edged	'Dominant Silver', 'Dominant Dilute'
ashen fallow	Ashen Fallow	'Recessive Silver' (Europe/USA)
bronze fallow	Bronze Fallow	'Fallow'
NSL ino	NSL Lutino	'Recessive Lutino'
yellowcheek	Yellowcheek	'Sex-linked Yellowcheek', 'Sex-linked Yellowface'
tangerine	Yellowface	'Dominant Yellowcheek', 'Dominant Yellowface'
?	'Emerald'	'USA Olive', 'Suffused Yellow'
?	'Goldcheek'	N/A
faded	Faded	'West Coast Silver', 'Recessive Cinnamon'
platinum	Platinum	'Cinnamon', 'East Coast Silver'
greywing	Dilute	'Pastel Silver', 'East Coast Silver'
edged dilute	Edged Dilute	'Silver Spangle', 'Dominant Silver'
?	'Australian Fallow'	'Fallow'
suffused	Suffused	'Australian Olive'
pewter	Pewter	N/A
?	'Australian Yellowface'	N/A

HEALTH AND DISEASE

Dominant Edged Pearl Cockatiel hen.

Disclaimer

This section is designed to assist bird owners to understand disease in their birds. It is not meant or designed to take the place of the person best qualified to assist you in the diagnosis and treatment of your bird—your avian veterinarian. Nor is it, by any means, a complete A–Z of Cockatiel diseases—it would take an entire book to cover all of these, not just a chapter. It cannot be stressed enough—if you have a sick bird, take it to an avian veterinarian. Pet shop and folklore remedies rarely work, and the time wasted while waiting for a response can be the difference between a live bird and a dead one.

Any treatments or procedures described in this book are meant to help you understand bird health better. The authors and publishers accept no responsibility or liability for any person using these treatments without direct veterinary advice.

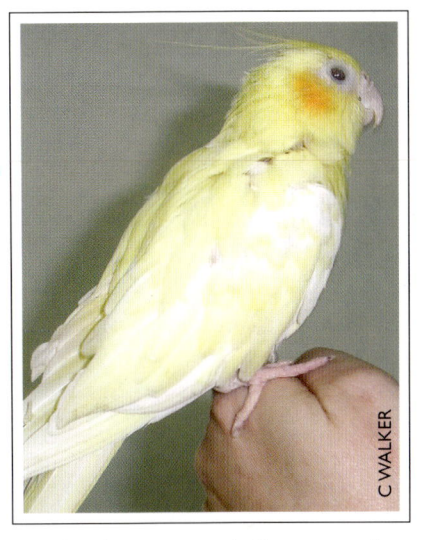

At the first sign of illness, realise that the bird is much sicker than it appears.

General Information

Cockatiels, as small parrots, can be found fairly close to the bottom of the food chain. In common with many other 'prey' species, evolutionary instincts have taught them that predators (eg falcons, snakes and quolls) prefer to single out the bird that looks slightly different—particularly if it looks sick and easy to catch. So one of their survival instincts is to not look sick.

One hundred years of domestication cannot cancel out millions of years of evolution. Our pet and aviary Cockatiels still retain the ability to hide signs of illness until they are very sick. To us, as bird keepers, this means that often the first signs of illness are seen at the end of the illness—not the beginning. The natural reaction to 'wait and see' can therefore be a death sentence for many sick Cockatiels.

You also need to be aware that sick Cockatiels will often show the same signs of disease—regardless of the cause. Nearly all sick birds will fluff their feathers, sit still and close their eyes. This is a response to disease, which is designed to conserve energy and body heat. It is not characteristic of any one disease—it simply means that the bird is sick.

As a bird owner you need to be alert for early signs of illness. Once seen and recognised, you then have to act on them quickly. The sooner a bird receives treatment, the better its chances of survival. As one veterinarian says, 'I cannot tell you what is wrong with your bird over the phone; all I can tell you is that the longer you leave it, the poorer the bird's chances are, and the higher your bill will be.'

How to Use This Chapter

This chapter is divided into two parts. After observing and examining your bird, write down the main symptoms or signs of disease that you are seeing. Look these up in Part One of this chapter. Accompanying each symptom or sign of illness is a short explanatory note to make sure that you are describing it correctly, as well as a list of possible diseases or causes for each symptom. The number in brackets after many of the listed diseases or causes refers to Part Two of this chapter, where you will be able to gather more information on the problem. For more detailed information you should refer to **A Guide to Basic Health and Disease in Birds** by Dr Mike Cannon, published by **ABK Publications**. There is not enough scope in a book like this to include full details on every aspect of bird health.

Supply of Prescription Medications

Each Australian state has a Health Department which, amongst other things, regulates the supply of prescription drugs (eg antibiotics, pain killers and sedatives) to both people and animals. These regulations make it very clear to veterinarians that the supply of these medications for the treatment of animals not under their direct care is illegal. 'Direct care' means that the veterinarian must have recently examined and diagnosed the animal in question for the disease problem for which the medication is required. **Put simply, this means that unless the veterinarian has seen your birds, for this disease, within the last few months, it is illegal for that veterinarian to dispense any prescription medications such as antibiotics.**

These regulations exist to protect both you and your birds. The indiscriminate use of drugs will eventually mean that many of these drugs will become useless. Then, when either you or your birds need that drug in an emergency, it may not work.

Asking your veterinarian to supply you with antibiotics for birds that he/she has not seen is asking your veterinarian to break the law and risk heavy fines and/or deregistration. Unless you are willing to guarantee payment of these fines and financially support your veterinarian while he/she is deregistered (and therefore unable to practise as a veterinarian), please do not ask your veterinarian to break the law in order to save you a consultation fee and some laboratory tests.

PART ONE

Symptom	Explanatory Notes	Possible Causes
Change in Droppings		
Diarrhoea	Diarrhoea indicates that the faecal portion of a dropping (usually green-brown and cylindrical) is watery and unformed.	• gastrointestinal disease (1.2) • intestinal parasites (1.3) • Psittacosis (1.1) • Megabacteria (1.4) • liver disease (2.3) • zinc and lead poisoning (2.4)
Change in colour of the faecal portion	Faeces are usually green to brown in colour.	• Black droppings usually indicate that the bird is not eating. • Very pale, bulky droppings may indicate that your bird is not digesting its food properly.
Very large faecal portion	Faeces are usually small. Large, soft, unformed droppings are abnormal.	• egg laying (2.5) • pancreatic disease (2.7)
Fresh blood in the droppings	Bright red blood should not be present.	• uterine disease (2.5) • cloacal disease (eg infection, papillomas, prolapse or cancer)
Whole seed in the droppings	Seed is usually ground up and digested in the stomach and intestines. It should not be passed whole in the droppings.	• gastrointestinal disease (1.2) • Megabacteria (1.4)

Symptom	Explanatory Notes	Possible Causes

Change in Droppings (continued)

Excessive urine	The urine is the liquid part of the dropping.	• kidney disease (2.2) • diabetes (2.6) • liver disease (1.1; 2.3) • zinc poisoning (2.4)
Change in colour of the urates	The urates are usually white and crystalline in appearance, not … • green • pink or red • yellow.	• liver disease (1.1; 2.3) • kidney disease (2.2) • lead poisoning (2.4) • not eating

Appetite and Thirst

Increased appetite	A Cockatiel will normally eat well in the morning and evening, and then 'graze' during the day.	• diabetes (2.6) • Megabacteria (1.4) • jaw injuries. (The bird is actually not eating, but rather is attempting to do so.) • pancreatic disease (2.7)
Decreased appetite	A Cockatiel will typically consume several teaspoons of food each day.	• non-specific sign of illness • mouth or jaw injuries
Increased thirst	A healthy Cockatiel should drink 5–10ml of water per day.	• lead or zinc poisoning (2.4) • kidney disease (2.2) • liver disease (2.3) • diabetes (2.6) • heat stress
Decreased thirst	A healthy Cockatiel should drink 5–10ml of water per day.	• non-specific sign of illness • sufficient water in greenfoods and vegetables
Weight Loss	A healthy Cockatiel weighs 80–100 grams	• non-specific sign of illness

Posture

Fluffed appearance, immobile, eyes closed, both legs on perch	The healthy bird is alert and interested in its environment. Sleeping birds will often fluff their feathers and close their eyes, but usually hold one foot up.	• non-specific sign of illness. (Note: some sick birds will make an effort to appear normal when being examined, but can rarely maintain this for more than a minute or two.)
Sitting on the floor of the cage, fluffed, immobile, eyes closed	This is the next step in the progression of any illness. Even moderately ill birds will prefer to sit on a perch.	• severe illness. Seek veterinary assistance immediately.

Symptom	Explanatory Notes	Possible Causes

Posture (continued)

Tail pointing down perpendicular to cage floor	The healthy bird's back is a straight line, at an angle of about 45° from the horizontal.	• spinal deformity • eggbinding (2.5)
Tail bobbing up and down in an exaggerated movement	Movement of the tail is usually only just noticeable.	• respiratory disease (1.5) • internal organ enlargement, cancer • eggbinding (2.5)
Head held down, wings held out, tail spread	The healthy bird holds itself upright, wings folded against the body.	• reproductively active hen—courting posture (2.5)
Sitting on the floor of the cage, body upright, perhaps panting	Normally birds will prefer to sit on a perch.	• leg injuries • eggbinding (2.5)
Unsteady on feet, fitting, paralysed, other neurological signs	The healthy bird holds itself upright, wings folded against the body. Normally birds will prefer to sit on a perch.	• weakness due to illness • lead poisoning (2.4) • head trauma • severe calcium deficiency (2.1) • brain lesions, eg cancer or infection

In Cockatiels, powder-down feathers are found in two bands over the top of each thigh and the sides of the abdomen. These specialised feathers continually grow (provided that the bird is well). As they grow they tend to wear, leading to the formation of the white powder ('bloom') that, when spread through the feathers, gives them their water-repellent and lustrous character. This process is facilitated through regular preening. This bird had accidentally had its head closed in a door, resulting in virtually the total loss of both the upper and lower beak. An inability to preen resulted in the ungroomed powder-down feathers growing into a long 'apron' seen here behind the bird's legs.

The result of feather plucking.

Symptom	Explanatory Notes	Possible Causes

Feathers and Skin

Symptom	Explanatory Notes	Possible Causes
Feathering gradually darkening or becoming 'greasy-looking'	Feathers should be of normal colour and appearance.	• liver disease (1.1; 2.3) • malnutrition (2.1)
Broken tail feathers	The tail feathers should be long and graceful.	• heavy falls, often associated with poorly done wing clips • cage mate aggression • malnutrition (2.1)
'Stress lines', ie horizontal breaks in the feather vane	The barbs and barbules should be uniform along the length of the feather.	• stress or illness at the time the feather was emerging through the skin. If severe, it could indicate generalised disease in the bird.
Long straw-like feathers over the thighs	These are the powder-down feathers, and are usually chewed down by the bird while grooming, releasing the powder that helps to keep the feathers clean.	• inability to groom properly. This could be because of illness, spinal abnormalities or wearing an Elizabethan collar to prevent feather plucking.
Bird appears to be very itchy; feathers are ragged looking	Although a bird will groom itself frequently, it should not appear to be uncomfortable or frantic.	• lice and mites • malnutrition (2.1) • skin infections
Feathers missing on the head	Birds are unable to groom or feather pluck themselves on the head.	• cage mate aggression • skin or feather disease
Skin is very flaky and dry looking	The skin should be smooth, thin and almost transparent.	• malnutrition (2.1) • lack of bathing opportunities
Feather plucking	Chewing at, or pulling feathers out, is an abnormal activity for birds.	• skin infections • underlying painful lesions, eg bone cancer • passive smoking • reproductive disease (2.5) • pancreatic disease (2.7) • *Giardia* (1.3) • psychological causes, eg fear and boredom, are rare in Cockatiels.

Symptom	Explanatory Notes	Possible Causes
Wings		
Blood on feathers	Blood on the wingtips or feather shafts.	• wingtip trauma • broken blood feather • trauma to bone, muscle or skin
Wing drooping	The wings are usually held high and tightly against the body. Both wings are usually symmetrical.	• broken bones • muscle damage • weakness • respiratory disease (1.5)
Lumps	Localised or diffused swellings anywhere on the wing.	• cancer • healing or healed broken bone • feather cyst
Wings held away from the body	The wings are usually held tightly against the body.	• heat stress • fear • aggression
Green discolouration	Skin and underlying muscle are discoloured light to dark green.	• bruising
Feet and Legs		
Limping	Unable to take weight on a leg.	• injury to bone, muscle or joints • bumblefoot (infection in the sole of the foot)
Swollen joints	Swelling localised to any or all of the joints.	• gout (2.2) • arthritis • cancer • infection
Missing nails or toes	There should be four toes—two pointing forward and two pointing backwards.	• aggressive cage mates • unsafe caging or cage furniture
Abnormally shaped or directed feet and legs	Feet or legs permanently deviated in an abnormal direction.	• calcium deficiency while growing (2.1, 2.8) • poorly healed broken bone
Legs pointing out to the sides	One or both legs pointing outwards from the body.	• splay leg (2.8) • fracture • slipped tendon. (The tendon 'slips' out of the groove on the back of the hock. This is associated with trauma, poor diet and sometimes genetics.)
Green discolouration	Skin and underlying muscle discoloured light to dark green.	• bruising

Symptom	Explanatory Notes	Possible Causes
Head		
Overgrown beak, sometimes with bruising present in the beak	A healthy Cockatiel beak is usually reasonably short.	• liver disease (1.1; 2.3) • Scaly Face Mite. (This is rare in Cockatiels.)
Beak twisted to the left or the right	The upper and lower beaks should be symmetrical and meet evenly.	• scissor beak
Upper beak inside the lower beak	The upper beak should come down over the front of the lower beak.	• prognathism (a genetic defect requiring treatment by an avian veterinarian)
Flakes of keratin on the beak	The keratin of the beak should be smooth.	• poor nutrition (2.1) • lack of an abrasive surface in the cage
White crusty lesions on the beak	The keratin of the beak should be smooth.	• Scaly Face Mite
Nares (nostrils) unequal in size	Both nares should be open, symmetrical in size and 1–2mm in diameter.	• chronic respiratory disease (1.5) • Psittacosis (1.1)
Nares blocked, or staining or matting of feathers above the nares	The nares should be open, and the feathers above them free of any discharge, stain or matting.	• chronic respiratory disease (1.5) • Psittacosis (1.1)
Feather loss around the eyes	Feathers should grow right up to the eyelids.	• conjunctivitis (1.5) • sinusitis (1.5) • Psittacosis (1.1)
Thickening of the conjunctiva (the mucous membrane lining the inside of the eyelid)	The conjunctiva should be barely visible.	• conjunctivitis (1.5) • sinusitis (1.5) • Psittacosis (1.1)
Discharge from the eyes	There should be no eye discharge at all.	• conjunctivitis (1.5) • sinusitis (1.5) • Psittacosis (1.1)
Swellings on the face, eye protruding	When viewed from the front and above, both sides of the head should be symmetrical.	• sinusitis (1.5) • cancer • Psittacosis (1.1)

Symptom	Explanatory Notes	Possible Causes
Head (continued)		
Bald patch behind the crest	In most Cockatiels the head is fully feathered.	• The bald patch can occur in the Lutino Cockatiel, however, quality breeding is attempting to reduce this incidence. • Feather picking by a cage mate.
Matting of the feathers on the cheekpatch	Cheekpatches cover the ears and should be clean at all times.	• ear infection
Matting of the feathers over the face and head	The feathers should be clean and well groomed.	• vomiting (1.2; 2.4)

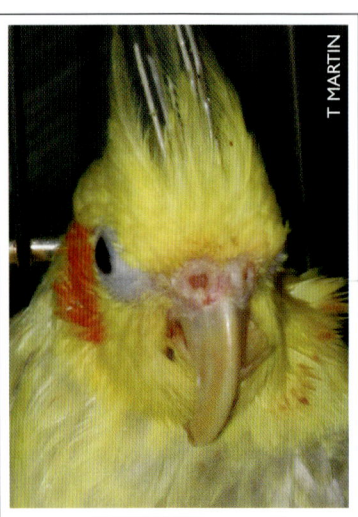

A healthy Cockatiel beak is reasonably short.

This Cockatiel displays an overgrown beak, a symptom of chronic liver disease.

Nasal exudate in a Cockatiel with Psittacosis (Chlamydiosis)—a common infection in Cockatiels that is not only potentially deadly to the bird, but can be transmitted to people.

Symptom	Explanatory Notes	Possible Causes
Physical Examination		
Prominent keel bone	When palpating the keel, the pectoral muscles should feel firm and form an angle of about 45° from the ridge of the keel. If the keel bone feels very thin, it indicates muscle wastage.	• chronic illness, causing catabolism—a breaking down of the muscle. • inability to fly, causing the muscles to shrink (degenerate) due to disuse. (Note: this degree of muscle wastage is usually not as severe as with illness.)
Twisted keel bone	The keel should run along a straight line from the chin to the vent.	• calcium deficiency (2.1) • genetic defect
'Cleavage' along keel	The pectoral muscles should not be so prominent as to form a cleavage along the keel.	• obesity (2.1)
Enlarged abdomen	The abdominal muscles, between the end of the keel and the pubic bones, should be concave. If they are convex, it indicates an enlarged abdomen pushing the muscles out. Note: enlargements within the abdomen can cause signs of respiratory distress and these birds must be handled very carefully.	• obesity (2.1) • enlarged liver (2.3) • cancer • reproductive disease in hens (2.5) • hernia
Lump on the back, at the base of the tail	This is the location of the preen gland. It is usually small enough that it is not noticeable.	• cancer • impaction of gland • infected gland
Air under the skin	The skin is usually tightly adhered to the underlying muscle. No air should be felt between the skin and the muscles.	• trauma • severe respiratory disease (1.5)

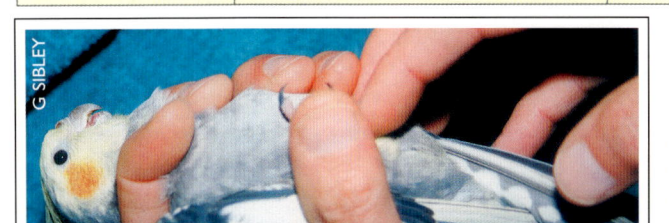

A method of restraint in order to perform a physical examination.

Paediatrics

Crop not emptying	The crop should be almost empty within 4–6 hours of being fed.	• bacterial or yeast infection in the crop (2.8) • food too hot or too cold (2.8) • food too thick or too watery (2.8) • generalised illness (2.8)
Air under the skin on the neck	There should be no air under the skin.	• ruptured air sac, often due to rough handling while being fed (2.8) • gulping air while feeding (2.8)
Skin very red	The skin should be pink.	• dehydration (2.8) • overheated (2.8) • illness (2.8)
Skin very pale	The skin should be pink.	• cold (2.8) • illness (2.8)
Overly large head	The head should be slightly larger in proportion to the rest of the body, but not too large.	• Stunting—reduced rate of growth due to any cause (2.8)
Bruising on the skin	There should not be any bruising on the skin.	• severe bacterial or viral infection (2.8) • injections, eg Baytril® (2.8)
Feathers not growing normally	The normal pattern of feather growth is wings, head and then body. The feathers should grow in straight lines along the feather tracts on the skin.	• Stunting—reduced rate of growth due to any cause (2.8)
Thin toes	Chicks should have uniformly fat toes.	• Stunting—reduced rate of growth due to any cause (2.8)
Vomiting	Occasionally a chick will regurgitate some food when its crop is over full, but otherwise there should not be any vomiting or regurgitation.	• crop infection (2.8) • generalised illness (2.8) • gut obstruction, eg from chewing on substrate (2.8)
Refusing to eat	Until ready to wean, chicks should be keen to eat when their crop is empty.	• weaning (2.8) • illness (2.8)
Reddened skin or scab over the crop	The skin over the crop should be pink.	• crop burn (2.8) • trauma from crop needle (2.8)

PART TWO

1. Infectious Diseases

1.1 *Psittacosis*

This disease is caused by a chlamydial organism *Chlamydophila psittaci*. Clinical signs most commonly seen in Cockatiels include conjunctivitis (swollen eyelids, reddened conjunctiva, ocular discharge and feather loss around the eye), sinusitis (swellings around the eye, nasal discharge, staining of feathers around the nares), sneezing, diarrhoea and green urates. However, clinical signs can be completely absent, or the bird may die suddenly without any clinical signs. Although most commonly seen in birds under stress (eg newly purchased; in pet shops; or overcrowded aviaries), Psittacosis can be seen in any bird at almost any time. There is a current trend amongst some aviculturists to say that 'all birds have this disease, so why worry about it?' The reason is very simple: **This disease is transmissible to people.** It causes flu-like symptoms, pneumonia and occasionally death. Seek veterinary advice immediately if your bird shows signs of this disease.

Psittacosis in a Cockatiel. Note the swollen red eyelids.

1.2 *Gastrointestinal Infections*

These infections can be due to a wide range of pathogens, including intestinal parasites, bacteria, yeast and viruses. Clinical signs can include vomiting, increased salivation, diarrhoea, whole seed in the droppings and weight loss. Death occurs due to dehydration, electrolyte loss and bacterial toxins damaging the liver and kidneys. Laboratory tests are usually required to determine the cause, and this then dictates the appropriate treatment. Affected birds must be rehydrated, usually by injections of fluids under the skin, although crop drenching with electrolytes is effective if the bird is not vomiting.

1.3 *Intestinal Parasites*

Intestinal parasites are, unfortunately, quite common in Cockatiels. Roundworms (ascarids) are commonly found in the intestinal tract. Their eggs can be identified under a microscope. *Cochlosoma* and *Giardia* are two motile (moving) protozoan (single-celled) parasites often found in the droppings

Vomiting in a Cockatiel. This can be due to a wide range of problems. In this case the bird was infected with a parasite, Trichomonas, that lives in the crop and causes severe inflammation and ulceration.

of Cockatiels. *Coccidia*, a non-motile protozoan parasite, is also occasionally found. All of these parasites can cause weight loss, diarrhoea, and even death. Diagnosis requires faecal examination under a microscope to detect either the eggs or the parasites themselves. Treatment is determined by the type of parasite detected. As these parasites are spread through the droppings of infected birds, it is important to minimise contact between birds and their droppings. Wire floor cages and good hygiene are excellent methods of minimising this contact. Water and feed dishes must be located so that birds cannot defecate into them.

1.4 Megabacteria

Despite its name, Megabacteria is a fungus found in the stomach of infected birds. Birds are infected when they eat food contaminated with the droppings of another infected bird. The fungus forms a dense mat on the stomach wall, preventing digestion of the food. Ulcers may also form. The infected bird becomes ravenously hungry, but loses weight. Blood may be seen around its mouth, and diarrhoea is not uncommon. Treatment requires fluids to rehydrate the bird, an antifungal drug (amphotericin B) by mouth for at least 10 days, and reduction of stress. Crop feeding with an easily digested food may be necessary in some cases.

1.5 *Respiratory Infections*

These can occur in the sinuses, airways, lungs or air sacs. Possible causes include bacterial infections, fungal infections (Aspergillosis), *Chlamydophila psittaci* (Psittacosis), inhaled food (eg handrearing formula or seed) and occasionally viral infections. Parasite infections are relatively rare. (Cockatiels are notorious for inhaling millet seeds into their trachea, causing a sudden obstruction.) Affected birds may sneeze, have discharges from the nares and eyes, swollen sinuses, open-mouth breathing and increased respiratory effort (wing droop, tail bobbing and exaggerated visible movement of the body with each breath). It must also be noted that eggbinding, enlarged internal organs and fluid in the abdomen can all put pressure on the air sacs and cause signs of respiratory distress—without actual respiratory disease being present. Veterinary attention should be sought immediately, as respiratory infections can be life threatening.

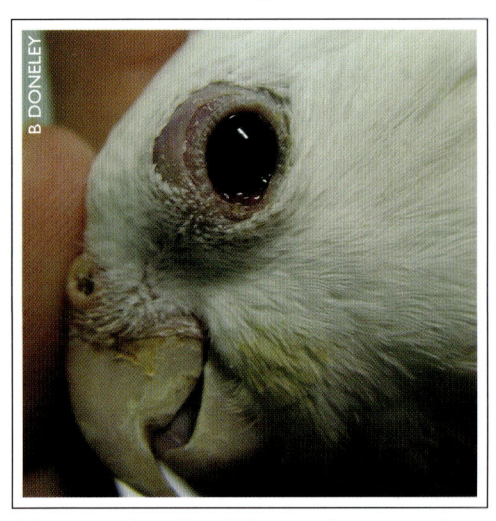

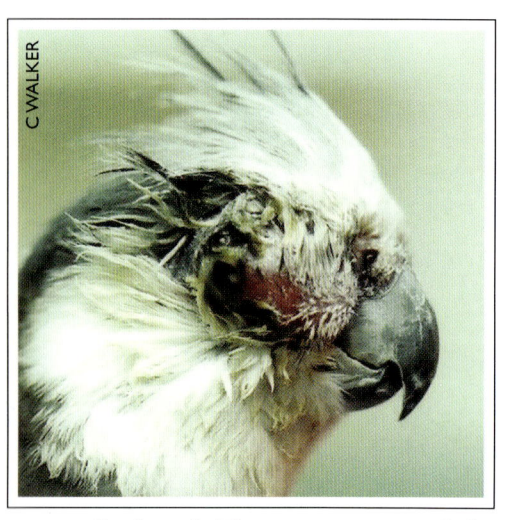

Above and right: Advanced sinus infection in a Cockatiel. The sinuses around the eyes interconnect via fine ducts that eventually drain into the back of the throat. With infection, inflammatory fluid can accumulate in the sinuses. If excessive, this cannot drain, leading to fluid-filled, bulging sinuses developing around the face. Under anaesthetic a bird's sinuses can be lanced and flushed.

1.6 *Psittacine Beak and Feather Disease (PBFD)* and *Avian Polyomavirus (APV)*

These are two of the viral diseases commonly found in other parrots, however are rarely seen in Cockatiels. This same situation is seen in other Australian cockatoos, and may reflect that these viruses have evolved alongside these parrots, developing a symbiotic relationship, ie the birds can carry the virus but do not develop disease. This symbiosis may mean that Cockatiels can carry these viruses, even though they do not show any outward signs of the disease, and may act as a source of infection for other (more susceptible) species. For this reason it cannot be assumed that they are free of these viruses without testing for them. You should speak to your avian veterinarian about the need for testing, and the best way to do so.

2. Non-Infectious Diseases

2.1 *Malnutrition*

This is one of the greatest problems confronting Cockatiels in captivity. An all-seed and, conversely, an all-pellet diet are causing more problems with health and longevity than infectious diseases and poor husbandry combined. It is a myth that Cockatiels are seed-eaters. In the wild they consume a wide range of foods. Ripened agricultural grain is only a relative newcomer to their diet. Seed is high in fat and carbohydrates, and low in virtually everything else, eg protein, vitamins and minerals. Birds fed this sort of diet develop the same problems that people eating high-fat diets do, including obesity, liver disease, diabetes and cardiovascular disease. On the other hand, all-pellet diets fed to colour mutation Cockatiels have been associated with kidney disease. A balance must be struck by feeding a mixture of pellets, vegetables and perhaps some seed. (See *Feeding* on page 35.) Your avian veterinarian will also be able to advise you on a healthy diet for your Cockatiel.

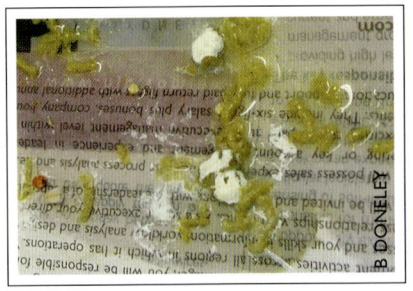

Above: Polyuria, or excessive urine in the droppings, can be due to kidney disease, diabetes or stress. Feeding an all-pellet diet to some Cockatiels can damage the bird's kidneys, leading to this problem.
Below: The colour changes in this bird's plumage, as well as the overgrown beak, are commonly seen in cases of chronic liver disease.

2.2 *Kidney Disease*

This can be due to a wide range of conditions—zinc and lead poisoning, bacterial infections and all-pellet diets, to name just a few. It is characterised by increased thirst and very watery droppings. If recognised early and treated aggressively, kidney disease can be reversed in many cases. Long-term kidney disease can lead to the development of gout—the deposition of uric acid into the joints—causing painful swellings of the joints and lameness. This condition is extremely difficult to treat, as only temporary pain relief is usually possible.

2.3 *Liver Disease*

Liver disease is also due to a wide range of conditions. In Cockatiels the two most common problems are Psittacosis and obesity (fatty liver). In acute cases affected birds often present with green urates, excessive urine in their droppings and non-specific signs of illness. Longer term cases often have darkened or greasy-looking feathers, an overgrown beak and weight loss. Fortunately the liver is a resilient organ. Once the problem is recognised and the cause identified, appropriate treatment usually results in complete recovery.

2.4 Zinc and Lead Poisoning

Metal poisoning is still far too common, especially in pet Cockatiels that are allowed to roam around the house and yard unsupervised. Both lead and zinc can be acute poisons, but lead can also be a cumulative toxin, ie a small amount taken into the body is stored, and when the levels are high enough, poisoning results. Both poisons will cause increased thirst, vomiting and watery droppings. Lead is also more likely to cause neurological signs such as fits and convulsions, paralysis and sometimes foot-chewing. Once again, prompt recognition and appropriate treatment are usually successful.

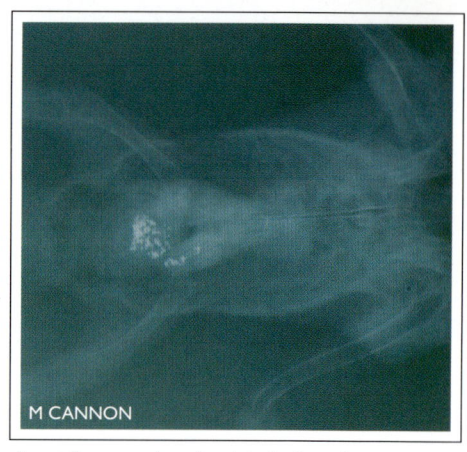

A radiograph of a bird that has chewed the galvanised wire on its aviary. The bright white dots in the abdomen are metal particles.

2.5 Reproductive Disease

This is a common consequence of the little Cockatiel's reproductive prowess. When combined with an inadequate diet (usually seed without any supplementation), the Cockatiel's ability to start egg laying at an early age and then keep on laying egg after egg is a recipe for disaster. Eggbinding, uterine infections, yolk peritonitis and cloacal prolapses are all common. Secondary problems such as pancreatitis, diabetes and liver disease are frequent consequences. Rarely a problem in aviary birds, these conditions are frequently fatal in pet Cockatiels. Continuous egg laying is an abnormal condition and needs to be treated aggressively. Changes in diet, environment and behaviour are the first steps, followed by hormonal therapy and even surgery. For more advice, you should consult your avian veterinarian.

This pet Cockatiel hen regularly laid eggs on the bottom of her cage. The paper substrate was removed and the egg laying stopped.

Continuous egg laying is a problem requiring immediate attention.

2.6 *Diabetes*

Diabetes is more common in Cockatiels than in any other bird species. It is more common in hens than cocks, possibly because of pancreatic damage due to yolk peritonitis. Obese birds are also more prone to this disease, possibly because of changes to the body's glucose metabolism. Insulin injections are not usually required, but supportive care while the underlying cause is corrected is necessary to treat this condition.

2.7 *Pancreatic Disease*

This disease is often the result of obesity or reproductive disease, but zinc poisoning and bacterial infections can be involved. Initially, birds with pancreatic disease may present with signs that include vomiting, diarrhoea, loss of appetite and plucking at their abdomen. As the condition progresses the pancreas becomes scarred and functions poorly. This can lead to either diabetes or poorly digested food (seen as large, pale bulky droppings). Early recognition and treatment are essential, as the chronic form is often impossible to reverse.

2.8 *Paediatric Problems*

Paediatric problems are commonly seen, and not just in handreared birds. When assessing paediatric problems in chicks, it is important to look at three important components:
- parental factors—the genetic make-up, diet and health of the parent birds;
- incubation—natural versus artificial; temperature, humidity, ventilation, turning, hygiene and record-keeping; and
- neonatal and paediatric factors—diet (composition and preparation); feeding routines (frequency, implements used, hygiene); quarantine (rearing other people's birds); and environment (temperature, humidity, privacy and security).

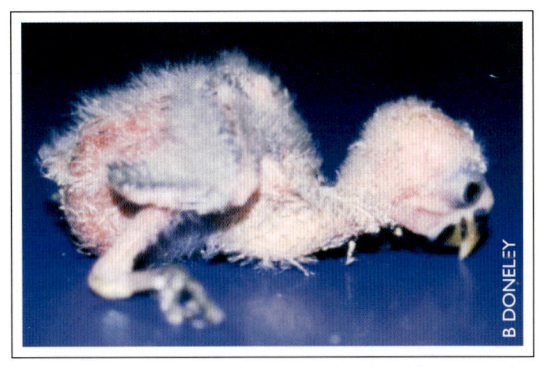

Stunting in a handreared chick. This can be due to any one of a number of problems that stunt the chick's growth. Until the original cause is identified and remedied, this chick will not improve. However, once the original problem is corrected, stunted chicks often recover fully and fledge at normal weights and times.

Chicks that are unwell, for any reason, are critical care patients. They have limited body reserves to draw on, and a poorly functional immune system. Key factors in a successful treatment include maintaining body temperature, correcting dehydration, eliminating the causative factor(s) and maintaining appropriate nutrition. Each of these factors must be considered for each patient, otherwise deterioration and death nearly always result. For more information, the reader is referred to ***A Guide to Incubation and Handraising Parrots*** by Phil Digney (published by **ABK Publications**).

Publisher's Note

This chapter is merely an introduction to avian medicine. Its purpose is to demonstrate the complexity of diagnosing and treating disease in Cockatiels. While veterinarians can appreciate that bird owners want to cut down on veterinary expenses, experience has shown that this is a short-sighted approach. Invariably, costs and losses are far greater when bird owners attempt to treat their birds without an accurate diagnosis and specific treatment. Although readers may wish to use this chapter to understand their birds' problems, it is strongly recommended that they consult an avian veterinarian if their birds are sick.

BIBLIOGRAPHY

Edged Dilute Cockatiel cock.

Adams, M, Baverstock, PR, Saunders, DA, Schodde, R & Smith, GT 1984, 'Biochemical systematics of the Australian cockatoos (Psittaciformes: Cacatuinae)', *Australian Journal of Zoology*, vol. 32, pp. 363–377.

Alderton, D 1989, *A Birdkeeper's Guide to Cockatiels*, Tetra Press, USA.

Barrett, G, Silcocks, A, Barry, S, Cunningham, R & Poulter, R 2003, *The New Atlas of Australian Birds*, Shannan Books, Australia.

Boles, WE 1993, 'A new cockatoo (Psittaciformes: Cacatuidae) from the Tertiary of Riversleigh, north-western Queensland, and an evaluation of rostral characters in the systematics of parrots', *Ibis*, vol. 135, pp. 8–18.

Brown, D 2003, *Under the Microscope—Microscope Use and Pathogen Identification in Birds and Reptiles*, ABK Publications, New South Wales.

Brown, DM & Toft, CA 1999, 'Molecular systematics and biogeography of the cockatoos (Psittaciformes: Cacatuidae)', *Auk*, vol. 116, pp. 141–157.

Cannon, MJ 2002, *A Guide to Basic Health and Disease in Birds*, rev. edn, ABK Publications, New South Wales.

Cooke, D & Cooke, F 1993, *Keeping and Breeding Cockatiels—A Complete Guide*, Blandford Press, UK.

Cross, P & Andersen, D 1994, *A Guide to Cockatiels and Their Mutations—Their Management, Care and Breeding*, Australian Birdkeeper, New South Wales.

Digney, P 1998, *A Guide to Incubation and Handraising Parrots*, ABK Publications, New South Wales.

Dixon, M 1994, 'Phylogenetic analysis of Psittaciformes (Aves)', PhD thesis, University of Texas, Austin.

Dorge, R & Sibley, G 1996, *A Guide to Pet and Companion Birds*, ABK Publications, New South Wales.

Forshaw, JM 2006, *Parrots of the World: An Identification Guide*, Princeton University Press, Princeton and Oxford, plate 6.

Forshaw, JM 1981, *Australian Parrots*, 2nd edn, Lansdowne Editions, Melbourne.

Forshaw, JM 1978, *Parrots of the World*, 2nd edn, Lansdowne Editions, Melbourne.

Hayward, J 1992, *The Manual of Colour Breeding*, The Aviculturist Publications.

Heidenreich, B 2005, *The Parrot Problem Solver—Finding Solutions to Aggressive Behaviour*, TFH Publications, USA.

Heidenreich, B 2004, *Good Bird! A Guide to Solving Behavioural Problems in Companion Parrots*, Avian Publications, USA.

Hesford, C 1998, 'The opaline factor in Australian parakeets', *The Genetics of Colour in Budgerigars and Related Parrots*, http://www/parrotgenetics.info.

Jardine, W 1836, *The Naturalist's Library*, vol. 10, WH Lizars, Edinburgh, plate 30.

Johnstone, RE & Storr, GM 1998, *Handbook of Western Australian Birds*, vol. 1, Western Australian Museum, Australia.

Jones, D 1987, 'Feeding ecology of the Cockatiel *Nymphicus hollandicus*, in a grain-growing area', *Australian Wildlife Research*, vol. 14, pp. 105–115.

Juniper, T & Parr, M 1998, *Parrots: A Guide to the Parrots of the World*, Pica Press, East Sussex.

Klasing, KC 1998, *Comparative Avian Nutrition*, CAB International, New York.

Koutsos, EA & Klasing, KC 2005, 'Vitamin A nutrition of growing Cockatiel chicks (*Nymphicus hollandicus*)', *Journal of Animal Physiology and Animal Nutrition*, vol. 89, pp. 379–387.

Koutsos, EA, Tell, LA, Woods, LW & Klasing, KC 2003, 'Adult Cockatiels (*Nymphicus hollandicus*) at maintenance are more sensitive to diets containing excess vitamin A than to vitamin A-deficient diets', *Journal of Nutrition*, vol. 133, pp. 1898–1902.

Madsen, CS, DeKloet, DH, Brooks, JE & DeKloet, R 1992, 'Highly repeated DNA sequences in birds: The structure and evolution of an abundant tandemly repeated 190-bp DNA fragment in parrots', *Genomics*, vol. 14, pp. 462–469.

Martin, S & Friedman, S 2004, *Training Animals: The Art of Science*, http://www.naturalencounters.com/papers/Training_Animals_The_Art_of_Science-Steve_Martin.pdf.

Martin, T 2002, *A Guide to Colour Mutations & Genetics in Parrots*, ABK Publications, New South Wales.

Martin, T 1999a, 'The Pied Red-rumped Parrot—a mysterious combination', in C Hesford's *The Genetics of Colour in Budgerigars and Related Parrots*, http://www/parrotgenetics.info.

Martin, T 1999b, 'Nature of the opaline locus', in C Hesford's *The Genetics of Colour in Budgerigars and Related Parrots*, http://www/parrotgenetics.info.

Martin, T 1999c, 'An update on the opaline locus', in C Hesford's *The Genetics of Colour in Budgerigars and Related Parrots*, http://www/parrotgenetics.info.

McGraw, K 2006, in *Bird Coloration—Mechanisms and Measurements*, vol. 1, K McGraw & G Hill (eds), Harvard University Press, USA.

McKendry, J 2007, 'Egg laying and aggression in pet Cockatiels', *Australian BirdKeeper Magazine*, vol. 19, no. 6, ABK Publications, New South Wales, pp. 363–365.

McKendry, J 2006, 'Regression in juvenile parrots', *Australian BirdKeeper Magazine*, vol. 19, no. 2, ABK Publications, New South Wales, pp. 92–95.

National Cockatiel Society, 1998, 'Interview with Margie Mason, developer of the new US Suffused Yellow mutation', *NCS Journal*, September/October issue, p. 32.

Ovenden, JR, MacKinley, AG & Crozier, RH 1987, 'Systematics and mitochondrial genome evolution of Australian rosellas', *Molecular Biological Evolution*, vol. 4, pp. 526–543.

Paull, G 1992, 'Silver Cockatiels', *Australian BirdKeeper Magazine*, vol. 5, no. 2, ABK Publications, New South Wales.

Simpson, K & Day, N 1986, *The Birds of Australia: A Book of Identification*, Lloyd O'Neil Pty Ltd, Dai Nippon Printing Co, Hong Kong.

Sindel, S 2003, *Australian Coral-billed Parrots (The Alisterus, Aprosmictus and Polytelis Genera)*, Singil Press, New South Wales.

Sindel, S & Lynn, R 1989, *Australian Cockatoos*, Singil Press, New South Wales.

Smith, GA 1978, *Encyclopedia of Cockatiels*, TFH Publications, USA.

Smith, GA 1975, 'Systematics of parrots', *Ibis*, vol. 117, pp. 18–68.

The Acclaimed 'A Guide to...' series

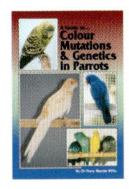

■ **A Guide to Colour Mutations & Genetics in Parrots**

The text is presented in three parts. The author takes the reader through a comfortable and absorbing introduction to understanding the basic principles of primary and combination mutations and colour genetics in the first two parts. Part three is for the adventurous, being the technical manifest of genetics in parrots. The text is supported throughout this 296-page title with over 700 colour photographs, numerous breeding expectations, an index of primary mutations in all parrot species, and glossary and explanation of genetic terms.
This highly readable and definitive work will become a major reference source internationally for many years to come.

Hard Cover   ISBN 978 0 9577024 7 9   Author: Dr Terry Martin BVSc
Soft Cover   ISBN 978 0 9577024 6 2

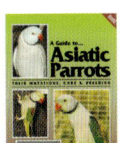

■ **A Guide to Asiatic Parrots (Revised Edition)**

This complete revision sees the addition of over 70 colour images of new mutations, improved text including new genetic tables and more information on nutrition, housing and breeding. Species include the Indian Ringnecked Parrot and their many mutations, the Alexandrine, Plum-headed, Derbyan, Malabar, Slaty-headed, Moustache, Malayan Long-tailed and Blossom-headed Parrot.

ISBN 978 0 9587102 5 1   Authors: Syd and Jack Smith

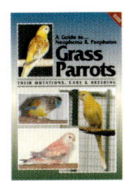

■ **A Guide to Neophema & Psephotus Grass Parrots (Revised Edition)**

One of ABK Publications' top selling *A Guide to...* titles worldwide, this revised edition features over 160 full colour images of mutations, examples of breeding expectations, housing, feeding and management. Species include the Bourke's, Turquoise, Scarlet-chested, Elegant, Blue-winged, Rock, Orange-bellied, Red-rumped, Mulga, Blue-bonnet, Hooded, Golden-shouldered and Paradise Parrot.

ISBN 978 0 9587102 4 4   Author: Toby Martin

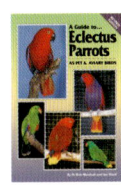

■ **A Guide to Eclectus Parrots (Revised Edition)**

This 160-page title features a comprehensive description of all 10 subspecies. Chapters include: Taxonomy and Identification, Eclectus in the Wild, Eclectus in Captivity—as Pet and Aviary Birds, Housing, Feeding, Breeding, Artificial Incubation and Handraising, Troubleshooting and Symptoms of Breeding Failure, Taming and Training, Colour Mutations and Genetics, Diseases and Disorders.
Featuring over 250 colour photographs, this 160-page title is available in both soft and hardcover format.

Hard Cover   ISBN 978 0 9750817 0 9   Authors: Dr Rob Marshall BVSc MACVSc (Avian Health)
Soft Cover   ISBN 978 0 9750817 0 9   and Ian Ward

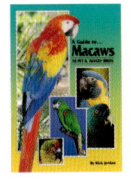

■ **A Guide to Macaws as Pet and Aviary Birds**

Featuring spectacular full colour photography throughout, this 136-page softcover title is packed with valuable and highly useable information.
Section 1: Breeding Macaws (including aviary design and construction, nutrition, breeding, incubation and handrearing, diseases and disorders).
Section 2: Macaws as Pet and Companion Birds (including housing, feeding, health aspects, behavioural problems).
Section 3: Species (with details on all macaw species).

ISBN 978 0 9577024 9 3   Author: Rick Jordan

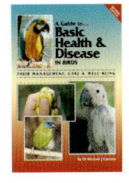

■ **A Guide to Basic Health & Disease in Birds (Revised Edition)**

The 112 pages of the book feature full colour photographs throughout, sketches, tables and easy-to-read valuable information. From finches to macaws, Dr Mike Cannon discusses common diseases, quarantine procedures, worming, parasite control, recognising health and nutritional problems, use of antibiotics, working with your avian veterinarian, and basic steps required to maintain healthy birds and good aviary management.

ISBN 978 0 9577024 5 5   Author: Dr Michael Cannon BVSc, MACVSc
                                             (Avian Health)

A Guide to Australian Grassfinches

Russell Kingston, a well-known aviculturist and finch breeder, has presented an informative and highly readable text that is lavishly supported with superb photographs. Each species is looked at individually and visual differences between the sexes are illustrated with some excellent sketches produced by the author. General topics covered include Acquiring Birds, Quarantine, Nutrition, Housing, Compatibility and Regulating the Breeding Season, to name just a few. A must for every finch breeder's library.

ISBN 978 0 9587102 2 0   Author: Russell Kingston

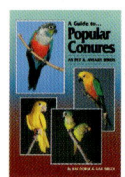

A Guide to Popular Conures

Combined with their own experiences, Ray Dorge and Gail Sibley have recorded the results of an extensive research of large conure breeders covering such topics as Conure Behaviour, Obtaining a Conure, Husbandry, Housing, Diet and Nutrition and Diseases and Disorders. Specific details on 13 of the most popular species kept in aviculture include Pet Quality, Terminology, Description, In the Wild and Breeding. Featuring beautiful full colour photography throughout, this title is sure to satisfy all fanciers and breeders of these parrots.

ISBN 978 0 9577024 3 1   Authors: Ray Dorge and Gail Sibley

A Guide to Australian White Cockatoos

Richly illustrated and full of practical hints, this well-researched book features facets of the author's personal experience which shine throughout its pages. Species featured are the Sulphur-crested Cockatoo, Short-billed Corella, Eastern Long-billed Corella, Major Mitchell's Cockatoo and Galah. Contents cover Management, Housing, Feeding and Nutrition, Breeding, Diseases and Disorders Common to Cockatoos and Species Profile covering Distribution, Subspecies, Sexing, Breeding, General Comments and Mutations.

ISBN 978 0 9577024 1 7   Author: Chris Hunt

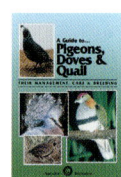

A Guide to Pigeons, Doves and Quail

Written by Dr Danny Brown, this book, a world first in aviculture, covers all species in this group available to the Australian aviculturist. Stunning colour photography throughout is supported by precise, easy-to-read information on the care, management, health and breeding of these unique birds.

ISBN 978 0 6462305 8 0   Author: Dr Danny Brown BSc (Hons) BVSc (Hons)
                                                                    MACVSc (Avian Health)

A Guide to Pet and Companion Birds

This informative and colourful 96-page book guides you through selecting a bird, housing, feeding and caring for that bird, understanding its behaviour, health aspects, taming, training, behavioural problems and how to increase your flock. A must for every pet bird owner.

ISBN 978 0 9587266 1 0   Authors: Ray Dorge and Gail Sibley

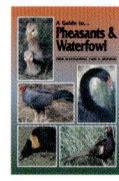

A Guide to Pheasants & Waterfowl

Author of the highly regarded *A Guide to Pigeons, Doves & Quail*, Dr Danny Brown has produced this superlative title on pheasants and waterfowl. The informative easy-to-read text is lavishly supported with beautiful colour images throughout. Covering all aspects of caring, housing, management and breeding of these unique birds, this title is a credit to the author and an ideal reference source.

ISBN 978 0 9587102 3 7   Author: Dr Danny Brown BSc (Hons) BVSc (Hons)
                                                                    MACVSc (Avian Health)

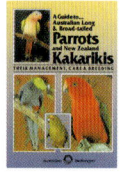

A Guide to Australian Long & Broad-tailed Parrots and New Zealand Kakarikis

Topics include the general management, housing, diet, care, breeding and handrearing of the Crimson-winged, Princess, Regent, Superb, King, Red-capped, Mallee Ringnecked, Cloncurry, Port Lincoln and Twenty-eight Parrot and the Red-fronted and Yellow-fronted Kakariki.

ISBN 978 0 9587455 3 6   Author: Kevin Wilson

■ **A Guide to Zebra Finches**

This title features 96 full colour pages of easy-to-read, highly informative text covering such topics as History and Ecology, Housing, Feeding, Health, Choosing Stock, Exhibiting and details of all currently recognised Australian colour varieties. A must for any Zebra Finch enthusiast.

ISBN 978 0 9577024 2 4   Authors: Milton, John and Joan Lewis

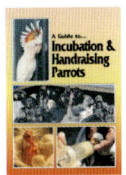

■ **A Guide to Incubation & Handraising Parrots**

Written by Western Australian aviculturist, Phil Digney, *A Guide to Incubation & Handraising Parrots* covers all necessary requirements needed to successfully take an egg through to a fully weaned chick. Beautifully illustrated with colour images throughout, this valuable title also includes many charts and diagrams and informative text laid out in an easy-to-read format. An invaluable reference for any serious bird breeder.

ISBN 978 0 9587102 1 3   Author: Phil Digney

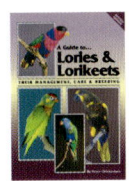

■ **A Guide to Lories & Lorikeets (Revised Edition)**

Completely reformatted and revised including new sections on Lories and Lorikeets as Pets, Diseases and Disorders and Colour Mutations and Breeding Expectations, this title is bigger (with 152 pages), better and more colourful than the highly successful original edition. Peter Odekerken's exceptional photography again is beautifully supportive of the informative text which together make this a must-have title.

ISBN 978 0 9577024 4 8   Author: Peter Odekerken

■ **A Guide to Gouldian Finches (Revised Edition)**

This complete revision features concise information on Gouldian Finches in the wild and in captivity, including Housing, Nutrition, Breeding, Mutation and Colour Breeding, Health and Disease. Supported by over 330 images, including an extensive selection of mutations.

ISBN 978 0 9750817 1 6   Edited by: ABK Publications
Contributing authors: Dr Milton Lewis BSc (Hons) PhD, Dr Terry Martin BVSc, Dr Rob Marshall BVSc MACVSc (Avian Health) and Ron Tristram

■ **A Guide to Black Cockatoos**

This title, featuring 300 colour photographs over 160 pages, is available in hard and soft cover.
Chapters on Black Cockatoos in Captivity include Taxonomy, Conservation, Housing, Nestlogs, Feeding, Breeding, Infertility, Nest Inspection, Artificial Incubation, Handrearing, Companion Birds and Diseases and Disorders.
The individual species section features all the black cockatoo species and subspecies of the genus *Calyptorhynchus*, the Palm Cockatoo *Probosciger aterrimus* and the Gang Gang Cockatoo *Callocephalon fimbriatum*.

Hard Cover ISBN 978 0 9750817 4 7   Authors: Neville and Enid Connors
Soft Cover ISBN 978 0 9750817 3 0

■ **A Guide to Grey Parrots**

This title authored by prolific avicultural author, Rosemary Low, stands alone when compared to many other books on the Grey Parrot. Rosemary presents interesting and diverse snippets gathered from various keepers' personal experiences in pet and aviary situations. Supported by numerous colour images, topics examine the Grey Parrot in the wild as well as their housing, management, feeding, nutritional requirements, behaviour, enrichment, breeding, handraising, training and veterinary aspects in captivity. A must-read for anyone contemplating to keep or currently keeping this hghly intelligent species.

Hard Cover ISBN 978 0 9750817 5 4   Author: Rosemary Low
Soft Cover   ISBN 978 0 9750817 6 1

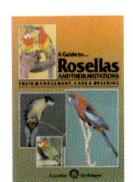

■ **A Guide to Rosellas and their Mutations (Under Revision)**

Simply the best publications on pet & aviary birds available ...

AUSTRALIAN BirdKeeper MAGAZINE

SUBSCRIPTIONS AVAILABLE

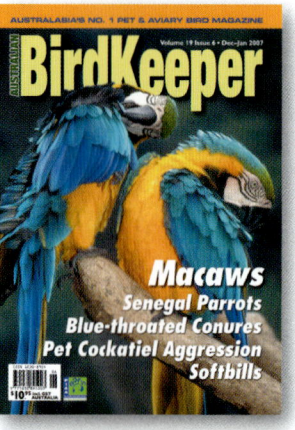

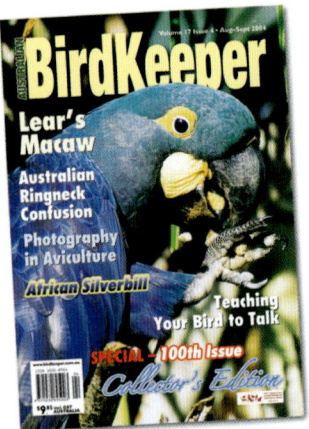

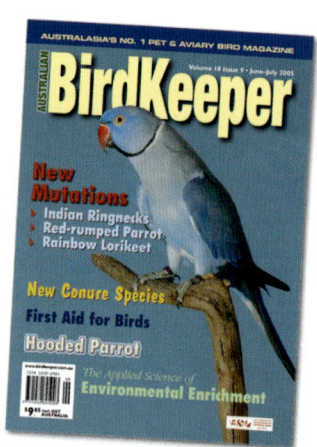

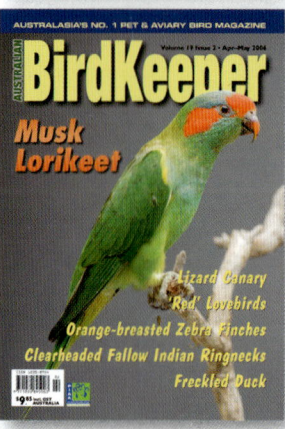

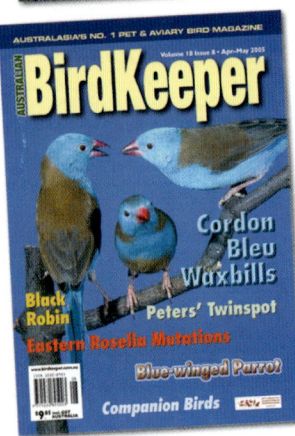

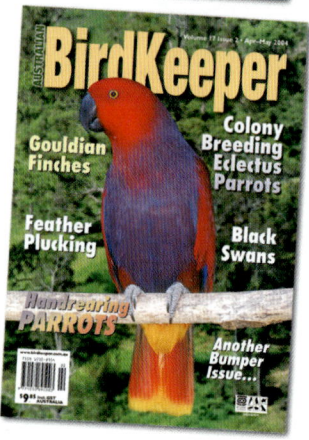

Six glossy, colourful and informative issues per year. Featuring articles written by top breeders, bird trainers, avian psychologists and avian veterinarians from all over the world.

For further information or free catalogue contact:

ABK PUBLICATIONS

PO Box 6288 South Tweed Heads NSW 2486 Australia
Phone: (07) 5590 7777 Fax: (07) 5590 7130
Email: birdkeeper@birdkeeper.com.au
Web site: www.birdkeeper.com.au